METAEPISTEMOLOGY
AND SKEPTICISM

Studies in Epistemology and Cognitive Theory
General Editor: Paul K. Moser, Loyola University of Chicago

METAEPISTEMOLOGY AND SKEPTICISM

Richard Fumerton

ROWMAN & LITTLEFIELD PUBLISHERS, INC.

ROWMAN & LITTLEFIELD PUBLISHERS, INC.

Published in the United States of America
by Rowman & Littlefield Publishers, Inc.
4720 Boston Way, Lanham, Maryland 20706

3 Henrietta Street
London WC2E 8LU, England

British Cataloging in Publication Information Available

Library of Congress Cataloging-in-Publication Data
Fumerton, Richard A.
Metaepistemology and skepticism / Richard Fumerton.
p.   cm.—(Studies in epistemology and cognitive theory)
Includes bibliographical references and index.
1. Knowledge, Theory of.   2. Skepticism.   I. Title.   II. Series.
BD161.F.F86 1995   121—dc20   95-32345   CIP

ISBN 0–8476–8106–8 (cloth: alk. paper)
ISBN 0–8476–8107–6 (pbk.: alk. paper)

Printed in the United States of America

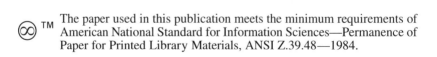

For Mom and Dad

Contents

Preface

In the last two decades a great many philosophers have accepted a revolutionary approach to the understanding of epistemic concepts, an approach that, if correct, should change the very way we think about the history and practice of epistemology. These philosophers seek to "naturalize" and "externalize" the concepts of epistemic justification, rationality, and knowledge and, in the course of doing so, they explicitly or implicitly suggest a new response to traditional skeptical concerns. If contemporary externalists are correct then most of the history of epistemology was radically misguided and confused.

While the internalism/externalism debate in epistemology has moved to center stage, I believe that we still lack a clear understanding of precisely what is *fundamentally* at issue between proponents of the respective views. A good part of my concern in this book is to define clearly the internalism/externalism controversy, or more precisely, the internalism/externalism controversies. I am particularly interested in exploring the implications of accepting various versions of internalism and externalism for the way in which one should understand and respond to the traditional skeptical challenges that over the years have so captured the imagination and attention of philosophers.

Although understanding the *connections* between metaepistemological views (views about the correct analysis of epistemic concepts) and approaches to skepticism is my primary concern, it will be obvious that I have my own axes to grind. I argue that there is a version of internalism that alone can understand epistemological questions in a way that makes their answers relevant to the kind of philosophical interest and curiosity that gives rise to the questions in the first place. Externalists may succeed in introducing interesting, clear, and perhaps, in some contexts, useful ways of understanding knowledge and justified belief. But we can accept this conclusion without conceding the *philosophical* relevance of these concepts. Convincing philosophers that their way of understanding concepts leaves them unable to ask questions they want to ask is obviously going to be an uphill battle. Paradigm internalists and externalists are firmly in their respective camps. The literature contains all sorts of counterexamples to philosophical analyses of epistemic concepts that succeed in convincing many to reject various positions. But the lines are pretty clearly drawn by now, and both internalists and externalists know what their views commit them to saying and they are usually willing to say it. If we are going to convince anyone that there is something wrong with paradigm versions of externalism, I think it may require a more subtle argument that reveals unacceptable consequences of the views when it comes to the way in which we would have to engage traditional skepticism. Ironically, I think the ease with which externalists can and *should* ignore skeptical challenges *at all levels* will eventually undermine for many the plausibility of the framework within which the externalist understands epistemological questions.

In the course of evaluating the consequences of my own views for skepticism, I argue that the alternatives for avoiding skepticism are stark. Although there is at least one dialectically attractive position that enables one to throw back the skeptical challenge, I have strong phenomenologically based reservations about its ultimate intelligibility. I shall try to convince the reader, however, that once one understands the source of philosophical interest in epistemology, one should not find unacceptable or even surprising the plausibility of even fairly radical skepticism.

In advancing these claims I sometimes paint with a very broad stroke. My concern is often not with the details of some particular externalist or internalist account of epistemic concepts but with what I take to be the fundamental presuppositions of the views. There probably has been as much philosophical work written on skepticism as on any other topic in philosophy. I have of necessity restricted my attention to certain paradigmatic representatives of views and traditions. There are many valuable and important books and articles that I do not discuss, and I hope my failure to do so is not interpreted as in any way disparaging their contribution to the field.

Acknowledgements

Many people have influenced my epistemological views over the years, and some of them are not explicitly discussed in the body of the book. Some of the ideas that find expression in the book were developed in professional talks and informal conversations with colleagues, and my ideas have often evolved as a result of comments and criticisms I received. I thank all of my colleagues at Iowa, and in particular four of our students, Eric Hockett, Baron Reed, Jim Sloan, and Paul Studtmann, who read early drafts of the manuscript and made a number of important comments and suggestions. I also thank Richard Foley, Ernie Sosa, Laurence BonJour, and Paul Moser, whose comments and advice on the manuscript, or work that found its way into the manuscript, were most valuable. Jan Kleinschmidt and Maurene Morgan have the patience of saints and helped me in countless ways. Finally, as always, I express my special appreciation to Patti, Tara, and Rob, who bring me so much happiness and who make everything I do in life so much easier as a result. Rob was also a great help with preparing the final draft of the manuscript.

Some of the issues addressed in chapter 5 were discussed in "The Incoherence of Coherence Theories," *Journal of Philosophical Research 19 (1994)*: 89–102, and some of the issues addressed in a section of chapter 4 were discussed in "Nozick's Epistemology," in *The Possibility of Knowledge,* edited by Steven Luper-Foy (Totowa, N.J.: Rowman & Littlefield, 1987). Also, some of the themes scattered throughout the book were explored in "Metaepistemology and Skepticism," in *Doubting,* eds. Michael Roth and Glenn Ross (Dordrecht: Kluwer Academic Publishers, 1990), 57–68. My views have changed significantly, however, since the publication of that article. Most of this book was written on a developmental leave provided by the University of Iowa.

Chapter One

Metaepistemology and Normative Epistemology

The Distinction Between Metaepistemology and Normative Epistemology

Although this terminology is relatively new, it has always been possible to distinguish two quite different sorts of questions in epistemology—metaepistemological questions and normative epistemological questions. The terminology is borrowed from, and the distinction in many ways parallels, the more familiar distinction in ethics between metaethics and normative ethics. Just as in metaethics one is primarily concerned with analyzing the concepts fundamental to ethical discourse, so in metaepistemology one is primarily concerned with analyzing the concepts fundamental to epistemological discourse. Just as in normative ethics one presupposes an understanding of ethical concepts and seeks to answer either general or specific questions concerning which actions are right, what things are good, and how people ought to behave, so in normative epistemology one presupposes an understanding of epistemological concepts and seeks to determine what one knows and rationally or justifiably believes.

I characterized metaepistemological questions as those *primarily* concerned with the analysis of epistemic concepts. We can also recognize *secondary* metaepistemological questions. We might be interested in discovering the relations between epistemic concepts and other concepts, and in discovering *properties* of epistemic concepts that are revealed through the analysis of such concepts but which should not be referred to in the analysis itself.[1] Thus, for example, the question of whether we always have the possibility of ''accessing'' our epistemic states *may* not need to be answered in the analysis of epistemic concepts even if we can discover the answer to that question solely through a

1

careful examination of the epistemic concepts we have analyzed. If questions about access can be answered without worrying about what, if anything, we actually do know or are justified in believing, we might still usefully call such questions metaepistemological. Again, the analogy with ethics is fruitful. The question of whether one can deduce an "ought" statement from an "is" statement is a metaethical question, but it might not be a narrow question about the analysis of ethical concepts. It might, however, be a question that one discovers the answer to through an analysis and examination of *concepts*, and as such it should be distinguished from questions in normative ethics.

Classical issues concerning skepticism seem to fall most naturally into normative epistemology. The arguments of the skeptic most often *presuppose* metaepistemological positions, but there is often far too little explicit discussion of the nature of knowledge or rational belief. Those of us who have always felt that metaethics has at least a logical priority over normative ethics are likely to feel that metaepistemology takes logical priority over normative issues in epistemology, and we may understandably feel that the way to get to the bottom of traditional skeptical problems is to arrive first at the correct metaepistemological positions. This view in epistemology is no more uncontroversial than its counterpart in ethics, and later in this chapter we will discuss the issue of whether metaepistemological investigation can take place in the absence of normative epistemological conclusions.

The importance of relating metaepistemological controversies to traditional issues in normative epistemology, particularly issues relating to the skeptical challenge, has never been more important. The internalism/externalism metaepistemological debate continues to rage, and its resolution has profound implications for the way in which philosophers should view issues in normative epistemology. Indeed, I believe that it is no exaggeration to suggest that if certain versions of externalism are correct, the history of epistemology is filled with philosophers who radically misconceived the nature of their enterprise. Moreover, if externalism is correct, the vast majority of contemporary philosophers are simply incompetent qua philosophers to address many of the questions that defined the history of western epistemology.

In what follows I am primarily interested in distinguishing a variety of metaepistemological views and understanding their implications for classical issues involving skepticism. I give this task priority in part because I am convinced that there simply are many different, even radically different, epistemological concepts that find expression in ordinary discourse. A search for *the* correct understanding of epistemic concepts is futile. However, I am far from neutral with respect to the sort of *interest* that an epistemic concept has for a philosopher, and I will try to convince you that there is a plausible version of internalism that captures the concepts of epistemology that have been presupposed by much of classical epistemology and that will continue to be presupposed by *philosophers* who are interested in a certain kind of question. I will also argue that this version of internalism may well lead to skepticism. That it does so, however, is no reason to reject it. Another of my primary concerns is to convince the reader that one must not take skepticism with respect to philosophically interesting normative epistemological questions as a reason to reject the metaepistemological views that contribute to that skepticism.

Metaepistemology

I have already characterized metaepistemology as first and foremost an investigation into the analysis of concepts fundamental to epistemological thought. Although talk about analyzing this and that is familiar to all of us in the analytic tradition, there never has been much metaphilosophical agreement on what analysis involves. There is also growing controversy over which concepts of epistemology are most fundamental. Let us examine this last question first, remembering that our metaepistemological investigations in this work are primarily concerned with shedding light on issues concerning skepticism.

Knowledge

Throughout the history of philosophy it is obvious that most philosophers concerned with epistemological questions raised those questions primarily using cognates of the verb "know" (or its synonyms). And indeed, the metaepistemological question "What is knowledge?" is at least as old as the *Theatetus*. In the history of philosophy, the skeptical conclusion is most often expressed as one about the limits of knowledge. One cannot *know*, the skeptic maintains, truths about the external world, other minds, the future, the past, the theoretical world of physics.

Despite the historical preoccupation with knowledge claims, however, I want to suggest that we focus our metaepistemological investigation, and subsequent discussion of the skeptical challenge, on the concept of justified or rational belief. My reasons are simple, albeit controversial. There is, I believe, a Cartesian conception of knowledge that always lurks beneath the surface and that continues to emerge in epistemological debates. When Descartes wanted knowledge, he wanted the absolute inconceivability of error. Descartes was convinced that he could withstand a skeptical challenge with respect to the possibility of getting this sort of knowledge, but few contemporary philosophers, myself included, think that there is any chance of defeating skepticism with respect to most commonsense knowledge claims if we employ a Cartesian conception of knowledge. Indeed, my own experience has been that the undergraduates we expect to be so outraged with the opening Cartesian skeptical worries of the *Meditations* soon respond with a shrug of the shoulders. So you cannot know with an *absolute certainty* that precludes any *conceivability* of error that there is a physical world. What did you expect? These kids were raised on *Star Trek*, where massive hallucination, mind transfers, and general sensory distortion form the plot lines of every second episode. In short, if we understand knowledge in terms of the inconceivability of error, the skeptical challenge becomes uninteresting because it is just too obvious who is going to win the debate.

Now this would not give most contemporary philosophers a reason for deemphasizing the importance of knowledge in the context of skeptical debate, because most contemporary philosophers will argue that the concept of knowledge is incorrectly understood as the inconceivability of error. It may well be that certain kinds of knowledge involve the *impossibility* of error, but even if one kind of impossibility should be explicated in terms of inconceivability, there are many other commonplace concepts of impossibility that have nothing to do with inconceivability. Thus an Armstrong can argue that basic knowledge involves the impossibility of mistake where this impossibility has something to do with lawful connections that hold between the fact that makes a belief true (and surrounding circumstances) and the occurrence of that belief.[2]

Furthermore, it is not clear to many that knowledge involves the impossibility of error in *any* sense of the term "impossibility." The famous "classical" conception of knowledge understands knowledge as justified true belief where the proponents of this view have usually emphasized that the kind of justification sufficient for knowledge need not logically or nomologically eliminate the possibility of error. As a number of philosophers have pointed out,[3] the "classical" conception of knowledge has a relatively short history and surprisingly few proponents, and it has fallen on hard times since the famous Gettier counterexamples. If the internalism/externalism debate is the focus of epistemology in the 1980s and 1990s, revising the justified true belief account of knowledge was the thriving industry of the 1970s. In no way do I wish to disparage the attempt to find a plausible analysis of knowledge that includes the concept of justification but does not rely on some notion of the impossibility of error. In fact, I do think that justified or rational true belief forms a core of one concept of knowledge that finds expression in ordinary usage, and I further believe that the key to resolving Gettier counterexamples is to expand the truth condition for knowledge, that is, to require something about the truth of propositions other than the proposition known. There is a plethora of such accounts. The most plausible seem to me to focus on either the truth of essential links in the edifice of justification or the unavailability of relevant true defeaters.[4] The latter may collapse into the former if one adopts a version of the view defended by Moser. According to Moser, in order to know that *P*, one must have a justified true belief that *P*, and there must be no undefeated defeater for that justification in the form of another true proposition. An undefeated defeater is a true proposition which, if it were added to one's justification, would destroy that justification when furthermore there are no other true propositions which, if they were added, would restore the *relevantly same* justification. Thus, if my justification for thinking that you are in Paris is that I received a phone call from you telling me of your location, the fact that somebody is going around saying that you are really in New York will not destroy my knowledge of your being in Paris, provided that the person is lying. The true proposition that the person is lying would negate the relevance of the misleading report were both propositions added to my evidence, and I could continue to employ my original inference in concluding that you were in Paris. Moser has a rather narrow understanding of what constitutes inferring a proposition in the same way, apparently focusing on deductive relations, and I would argue for a somewhat broader understanding, at least broad enough to recognize the relevance of nondeductive patterns of reasoning.[5]

As I suggested, an account like Moser's may collapse into an account that stresses the importance of having no essential falsehoods in the edifice of justification, provided that one can establish that whenever one has a justified belief, one must be implicitly justified in believing the falsehood of what Moser would call a relevant defeater, where that justified false belief forms an essential backdrop of justified presuppositions included in the entire edifice of justification for the belief that constitutes knowledge. So if one could make good the claim that when I infer that you are in Paris from our telephone conversation I also implicitly conclude that there are no sincere people testifying to the contrary, and if there is at least a sense in which this other (perhaps only dispositional) belief enters into the complete story of what justifies my belief, then the "no essential falsehood" and "no undefeated defeater" accounts might seem to give us the same result in this case. Although one cannot generalize from a single case, it does seem to me that, minimally, the two accounts will almost always give us the same result.

Now while there are in fact many non-Cartesian conceptions of knowledge floating around in ordinary discourse, I think it would be a mistake to conclude that the term "know" is *never* used in something like Descartes's sense. In fact, depending on the stakes, it seems to me that we sometimes come very close to using "know" to characterize the inconceivability of error. The familiar lottery problems are extremely difficult to accommodate with a non-Cartesian conception of knowledge. No matter how big the lottery, no matter how great the odds against winning, there is something very odd about claiming to know in advance the outcome of a fair drawing. Of course one need not turn to a concept of knowledge as the *inconceivability* of error to accommodate our intuition about the lottery example. With an analysis of knowledge that stresses merely the causal impossibility of error, one can also account for this intuition. But the proponents of such a view face an obvious dilemma. They want to be able to claim that we do know commonplace truths about the world around us based on perception and memory, for example. But there seems to be at least a statistical probability of perception or memory playing us false. The obvious move is to stress the particular circumstances surrounding this or that perception or memory. The claim is that even if it is not lawfully necessary that perceptual experiences with this qualitative character are veridical, it may be lawfully necessary that perceptual experiences with this qualitative character *in these circumstances* are veridical. But now one comes perilously close to losing the explanation of why we are reluctant to claim knowledge in the lottery cases. For all I know, the universe really is deterministic (at least at the macrolevel), and the outcome of the lottery was lawfully determined by a *complete* description of the antecedent state of the universe. Does the determinacy or indeterminacy of the universe seem the slightest bit relevant to the appropriateness of our claiming to know the outcome of that lottery?

I think a more plausible approach to the lottery problem is simply to concede that in the context of a knowledge claim about the outcome of a lottery, the person making such a claim is naturally presumed to have moved to a very strong conception of knowledge, requiring something very much like the inconceivability of error. After all, everyone knows (roughly) the probabilities, so if someone bothers to point out that one does not know that a given ticket will lose, what could be meant but that it is still entirely imaginable that the ticket will win? The imaginability of error with respect to the rest of our commonplace beliefs about the past, the future, and the external world is perhaps less vivid to those who have not read any philosophy, and when knowledge claims are made or denied with respect to more mundane matters it is understandable that it is a (statistically) more likely kind of error that people are denying or emphasizing.

I want to stress that I am not concerned very much in this book with knowledge as the inconceivability of error, and however hard one tries to make this clear, it seems to me that there is a danger, particularly when one discusses skeptical scenarios, that the reader will fall back into invoking some very strong conception of knowledge in reaching epistemological conclusions. Moreover, I think it will become clear that on the modified accounts of knowledge as justified or rational true beliefs, it really is justification or rationality that becomes the focus of philosophical interest from the first person perspective.[6] On defeasibility analyses of knowledge, for example, one is pretty much finished with one's epistemological investigation when one concludes that one is justified in believing *P*. Whenever one is justified in believing *P*, it will turn out that one is justified in believing that there are no undefeated defeaters. I realize that this has been denied by a number of philosophers, but their arguments strike me as entirely unconvincing. Dretske, for example, argued that when I see what looks like a zebra at the zoo I am perfectly justified in concluding that it is a zebra, in spite of the fact that I have no justification for thinking that someone has not carefully painted a mule to look like a zebra.[7] It is true that I may not have even contemplated the proposition that someone painted a mule to look like a zebra, but it seems to me that if I do not have available to me justification for ruling out that hypothesis as unlikely, I simply have no justification for thinking that the animal really is a zebra. And of course such justification (assuming a commonsense stance rejecting skepticism) seems readily available. This is not to deny, of course, that there is a difference between knowledge and justified belief. It is only to emphasize that once one has reached one's conclusions about what one is justified in believing, there is nothing else to do by way of ensuring that one has knowledge, on a justified true belief account of knowledge where the justification need not rule out the possibility of error.

Justified or Rational Belief and the Normativity of Epistemic Judgments

The metaepistemological question with which I am primarily concerned asks about the nature of justified or rational belief. The skepticisms I am interested in discussing are those that challenge our ability to have justified or rational beliefs with respect to a given kind of proposition. You may have noticed by now that I have most often disjoined the expressions "justified" and "rational" whenever I characterized the kind of belief in which I am interested. The primary reason for this involves a dispute that has developed over the concept of justification. Philosophers as diverse as Foley, Plantinga, Chisholm, Goldman, and Sosa all seem to think that there is a normative dimension to justification. And of course, I have encouraged precisely this way of talking by distinguishing between metaepistemology and *normative* epistemology. At least some of these philosophers, however, argue that one needs a concept other than the *normative* concept of justification to fill the role of that which must be added to true belief in order to get knowledge. I discuss competing philosophical accounts about the nature of justified or rational belief in detail in subsequent chapters. Still I think one can make a number of useful preliminary distinctions that will not presuppose answers to fundamental metaepistemological questions. Specifically, I think it is useful to distinguish a number of different concepts of justification and rationality and to discuss in a very general way the sense in which it is appropriate to regard *epistemic* justification and rationality as normative.

I might begin by emphasizing the contributions Plantinga in particular has made to the questions I address.[8] It is probably true of many philosophers (certainly of myself) that we begin using a locution like "justified belief" without being sensitive enough to its etymology and the implications of that etymology. As I have already noted in my brief discussion of knowledge, the concept of justified belief seems conspicuous by its absence in the history of epistemology. A. J. Ayer is almost always listed as one of the early proponents of a justified true belief account of knowledge, and it is interesting that his account of justification was expressed using language from ethical or moral theory. In order to know, according to Ayer, one must not only be sure but also have the *right* to be sure.[9] Ever since he first introduced the idea in *Perceiving*, Chisholm has continued to flirt periodically with the idea of analyzing epistemic concepts like justification using the concepts of value theory.[10] And it is simply uncontroversial that when contemporary epistemologists are discussing epistemological questions they often end up formulating those questions using the very terms that one uses in the formulation of ethical questions. The epistemologist asks what one *ought* to believe given that one has such and such evidence. To *praise* a belief as justified is to praise the person for believing what they ought to have believed (or at least for doing nothing *wrong* in holding the belief). What could be more natural than the conclusion that the concepts of justified or rational belief are normative concepts in exactly the same sense in which the concept of justified action, for example, is a normative concept? But in precisely what sense are epistemic concepts normative? What is involved in claiming that a concept is normative? How close is the analogy between ethics and epistemology?

In the first place, the epistemologist would do well to remember that an attempt to clarify metaepistemic controversies by relating epistemic concepts to ethical concepts might be like an attempt to clarify the writing of Hegel by relating it to the work of Heidegger. There is no more agreement on what the normativity of ethical judgments amounts to than there is even implicit agreement on what the normativity of epistemological judgments amounts to. Some ethical theorists contrast normative ethical judgments with descriptive judgments. The normativity of ethical judgments consists in the fact that they prescribe rather than describe features of the world. The paradigm of such a view is Hare's contention that at least some moral judgments are disguised imperatives. If being normative is contrasted with being descriptive, I suspect that there are relatively few epistemologists who wish to associate themselves with the view that claims about the justification or rationality of holding beliefs is normative.[11] On the other hand, a great many ethical theorists have taken normative ethical judgments to be a species of descriptive judgments. The adjective "normative" simply describes the subject matter of this sort of statement as distinct from some other sort of statement. How shall we begin an attempt to arrive at a neutral characterization of normativity without relying on the paradigm of ethical judgments and the details of some controversial metaethical position?

I suppose we could say that a concept *X* is normative if it can be defined or explicated[12] in part using value terms. We can then define value terms by simply listing a few of the most fundamental value terms and allowing as a value term any that can be defined in part using these. Thus we could say that a value term is "good," "bad," "ought," "should," "right," "wrong," or any term that can be defined in part using these. "Murder" might be a value term if murder should be understood as killing that is wrong. Is the concept of justified or rational belief normative in this sense? Well, I have already pointed out that epistemologists often raise their questions in terms of what one ought to believe given a certain body of evidence. So there is prima facie evidence that there is at least some sense in which the concept of justified belief is normative. But at this point we must be very careful in reaching any conclusions about the relationship between epistemology and ethics. Many philosophers would argue that the concept of what one ought to do or believe is multiply ambiguous. Some would argue that one must distinguish the "ought" of morality from the "ought" of prudence or practical rationality. And there may be no reason to stop here. There is the "ought" of etiquette, which may have nothing to do with what one morally or prudentially ought to do. And there is the "ought" of law, where again what one legally ought to do might be entirely irrelevant to what one morally ought to do and to what one prudentially ought to do. Finally, of course, there is the question of whether one must distinguish the epistemic "ought" from all of the above. There is strong reason to suppose that even when one is talking about what one ought to believe, one must distinguish the question of what one epistemically ought to believe from the question of what one morally ought to believe, or prudentially ought to believe, or perhaps even legally ought to believe (if one lives in a society odd enough to attempt to legislate belief). The same ambiguities exist with respect to the different senses of reasons for believing something. It seems that we can distinguish at least between moral, prudential, legal, and *epistemic* reasons for believing something.

Whether one can raise questions about the morality or prudence of having a certain belief is controversial in part because of notoriously difficult questions concerning the extent to which beliefs are the sorts of things under our control. Some would argue that it is only true that one morally ought to do X if X is something one can control, and that belief is not the kind of thing that is subject to our will.[13] Almost everyone will agree, however, that one can do things to affect one's belief and that there is at least a derivative sense in which one ought to believe, for example, that one will get well when one is sick, and believing that one will get well increases the chances of doing so. Acting on this obligation might consist only in doing what one can to maximize the chances of bringing about the belief. If one can distinguish the question of whether one morally ought to believe in the innocence of one's best friend, the question of whether one prudentially ought to believe in the innocence of one's best friend, and the question of whether one epistemically ought to believe in the innocence of one's best friend, then the suggestion that epistemic concepts are normative because what one is justified in believing is a function of what one ought to believe has a hollow ring to it. Presumably with the above ambiguities firmly in mind, it would only be plausible to claim that what one is epistemically justified in believing or what one is epistemically rational in believing is a function of what one epistemically ought to believe. But is the concept of what one epistemically ought to believe any more informative than the concept of what one would be epistemically rational or justified in believing?

The thesis that epistemic judgments are normative and the attempt to illuminate epistemic judgments by appealing to analogies with ethics might still be useful if one can find an important link between the "ought's" of morality, prudence, and epistemology. Foley and others have suggested that we do just that.[14] Crudely put, their idea is that normative judgments all relate to the efficacy of achieving *goals* or *ends*. There are different kinds of normative judgments concerning what we ought to do and what we ought to believe because there are different goals or ends that we are concerned to emphasize. Thus when we are talking about morally justified action or what we morally ought to do, the relevant goal is something like producing moral goodness (avoiding evil) and the actions that we ought to perform are those that are conducive to the goal of producing the morally best world. When we are concerned with what prudence dictates, however, the relevant goals or ends to be considered expand, perhaps to include everything that is desired intrinsically, for example. On one (rather crude) view, what one prudentially ought to do is what maximizes satisfaction of one's desires. What one ought to do legally or what one is legally justi-fied in doing is a function of the extent to which an action satisfies[15] the goal of following the law. What one ought to do from the standpoint of etiquette is a function of following the goals or ends set down by the "experts" who worry about such things. So all one has to do in order to fit the epistemic "ought" into this framework (and thus classify use-fully the kind of normativity epistemic judgments have) is delineate the relevant goals or ends that define what one epistemically ought to be-lieve. And the obvious candidates are the dual goals of believing what is true and avoiding belief in what is false.

If Pascal were right about his famous wager, belief in God might be the path one *prudentially* ought to follow, focusing on such goals as avoiding pain and seeking comfort. If you have promised your parents to believe in God, if it is good to keep a promise, and if there are no other good or bad effects of such a belief to consider, it might follow that prima facie you *morally* ought to believe in the existence of God. But neither of these normative judgments is relevant to whether you *epistemically* ought to believe in the existence of God. The only consideration relevant to this normative judgment is the efficacy with which such a belief contributes to the goals of believing what is true and avoiding belief in what is false.

Now as plausible and potentially illuminating as this account might seem initially, it is, I think, fatally flawed. In the first place, it must be immediately qualified to accommodate certain obvious objections. Suppose, for example, that I am a scientist interested in getting a grant from a religious organization. Although I think that belief in the existence of God is manifestly irrational (from the epistemic perspective), I discover that this organization will give me the grant only if it concludes that I am religious. I further have reason to believe that I am such a terrible liar that unless I actually get myself to believe in the existence of God they will discover that I am an atheist. Given all this *and my desire to pursue truth and avoid falsehood*, which I am convinced the grant will greatly enable me to satisfy, I may conclude that I ought to believe in the existence of God (or do what I can to bring it about that I believe in the existence of God). Yet by hypothesis this belief is one that I viewed as epistemically irrational. We cannot understand epistemic rationality simply in terms of actions designed to satisfy the goals of believing what is true and avoiding belief in what is false.

How might one modify the account to circumvent this difficulty? Foley suggests restricting the relevant epistemic goal to that of *now* believing what is true and *now* avoiding belief in what is false.[16] Even this, however, will fall prey to a revised (albeit more farfetched) version of the objection presented above. Suppose, to make it simple, that belief is under one's voluntary control and that I know that there is an all powerful being who will immediately cause me to believe massive falsehood *now* unless I accept the epistemically irrational conclusion that there are unicorns. It would seem that to accomplish the goal of believing what is true and avoiding belief in what is false *now*, I must again adopt an epistemically irrational belief.

The obvious solution at this point is to restrict the relevant goal that defines the epistemic "ought" to that of believing what is true now with respect to a given proposition. If I epistemically ought to believe that there is a God, the only relevant goal is that of believing what is true with respect to the question of whether there is or is not a God. If we say this, however, we must be very careful lest our account collapse the distinction between true belief and epistemically justified or rational belief. If we are actual consequence consequentialists[17] and we take what we ought to do or believe to be a function of the extent to which our actions and beliefs *actually* satisfy the relevant goals, then trivially we epistemically ought to believe in God when there is a God and we epistemically ought not believe in God when there is no God. Foley suggests at this point that it is something about beliefs an agent has, or more precisely would have after a certain process of reflection, about the efficacy of achieving the epistemic goals that is relevant to evaluating what one epistemically ought to believe. But there is a much more natural way of explicating the relationship between epistemic goals and what a person ought to believe, just as there is a more natural way of explicating the relevant relation that holds between a person's moral goals and what a person morally ought to do and a person's prudential goals and what a person prudentially ought to do.

The obvious move is to say simply that what a person ought to believe is a function of what that person is justified in believing would accomplish the goal of believing now what is true with respect to a given proposition. But that is, of course, a convoluted way of saying that what a person is justified in believing is what a person is justified in believing, an account entirely plausible but less than enlightening. Notice again that on many standard consequentialist accounts of morality or practical rationality, it is also crucial to introduce *epistemic* concepts into the analyses of what one morally or prudentially ought to do. I have argued in some detail that the concepts of what one morally ought to do and what one rationally ought to do are extraordinarily ambiguous. Although there are actual consequence consequentialist analyses of what one morally or rationally ought to do that find *occasional* expression in ordinary discourse, they are far from dominant. Consider the sadist who kills for pleasure the pedestrian in the mall when that pedestrian (unbeknownst to the sadist) was a terrorist about to blow up the city. There is surely a clear *sense* in which the sadist did not behave as he morally ought to have behaved.[18] The conventional poker wisdom that one should not draw to fill an inside straight is not falsified by the fact that this person would have filled the straight and won a great deal of money. How can we acknowledge that a person did what he ought to have done even when the consequences are much worse than would have resulted from an alternative? How can we acknowledge that a person behaved as he should not have behaved even when the consequences are far better than would have resulted from some alternative? The answer seems obvious. We must recognize the relevance of the epistemic perspective of the agent.

To determine what someone (morally or prudentially) ought to have done, we must consider what that person was epistemically justified in believing the probable and possible consequences of the action to be. Indeed, I have argued that there are literally indefinitely many derivative concepts of morality and rationality that also take into account what a person was epistemically justified in believing about the morality or rationality of actions, given more fundamental concepts of morality and rationality.[19] But if the analysis of familiar concepts of what a person ought to do must take into account the epistemic situation of the agent, it is simply a mistake to try to assimilate the epistemic ''ought'' to the ''ought'' of morality or practical rationality. In fact, an understanding of the ''ought's'' of morality and practical rationality is *parasitic* on an understanding of rational or justified belief. It would be folly, needless to say, to try to understand fundamental epistemic concepts in terms of what the agent was epistemically justified in believing about the probable and possible consequences of having a certain belief. Even philosophers who do not mind ''big'' circles in their philosophical theories will get dizzy traveling the circumference of this one.

So far the only sense in which we have acknowledged that epistemic judgments are normative is that they are sometimes expressed using an ''ought.'' That ''ought'' has been shown not only to be distinct from other ''ought's'' used in the expression of paradigm value judgments, but it has been shown to be *fundamentally* different. Nevertheless, we have not yet exhausted attempts to explicate normativity in a way that allows us to fit both epistemic judgments and our paradigm normative moral judgments under the same umbrella. It is sometimes claimed that our epistemic judgments are normative in that they implicitly involve *praise* or *blame* and *criticism*. Should we construe this as the relevant mark of normativity? Almost surely not. The problems with doing so are enormous. For one thing, however we define normativity, we want our paradigm of normative judgments, moral judgments, to fall under the concept. But it is far from clear what the relationship is between judging that someone did not do what he or she ought to have done and blaming or criticizing that person.

If you see a fire in the house next door and heroically attempt to save the people inside, I may conclude that you ought to have called the fire department instead of trying to solve the problem on your own. At the same time I might not *blame* you for failing to make the call. I might decide that under the circumstances it is perfectly natural for a person to panic and fail to do the rational thing. I might also think that you are just too stupid to figure out what you ought to do, and indeed, I might seldom blame you for the many idiotic things you do that you should not do. In short, there seems to be no conceptual connection between the evaluation of an agent's action and the praise or blame of the agent who acted that way. And if this seems right concerning the evaluation of what a person ought to have done, it seems even more obvious in the epistemic evaluation of a person's belief. Do we blame or criticize very stupid people for believing what they have no good epistemic reason to believe?[20] At the very least, logic does not require us to blame people for believing what it is epistemically irrational for them to believe.

It might be argued, however, that I am confusing the praise or blame of an agent with the positive evaluation or criticism of the agent's action or belief. "I am not criticizing you," someone might say, "I am criticizing what you did." And there surely does seem to be some sense in which when one's beliefs are called unjustified or irrational, one takes those *beliefs* to have been criticized. Shall we say that judgments about the epistemic justifiability or rationality of a belief are normative in that they imply praise or criticism of the *belief* (as opposed to the subject who has that belief)?

This is not helpful for two reasons. First, the notion of implying praise or criticism is simply too vague. When I tell the store owner that the knife I bought is extremely dull, there is surely a sense in which I am criticizing the knife (or implying criticism). When after test driving the car, I complain that it accelerates very slowly and pulls to the left, I am in some sense criticizing the car. But does that make "dull," "accelerating slowly," and "pulling to the left" normative expressions? Surely not. But why? One answer might be that there is no *conceptual* connection between judging that something has these characteristics and criticism. I might have wanted a dull knife to minimize the possibility of accident, for example. Now is there any *conceptual* connection between judging of a belief that it is epistemically irrational and criticizing the belief? Can we not imagine societies in which one values a kind of irrationality much the way a few people value dull knives? Indeed, I can think of a few philosophical movements that for all the world seem to place a premium on the incoherence of belief systems. And if that suggestion seems a little snide, can we not at least find some subculture of poets who explicitly disdain the confines of epistemically rational belief systems, the pursuit of truth, and so on? I have already agreed, of course, that there is a sense of "ought" that is customarily used in describing beliefs that a person is justified or rational in holding. And one can claim that if a belief is judged to be irrational it is being implicitly criticized as one that the subject ought not to have, but this will now take us full circle to the earlier problematic attempt to characterize the normativity of epistemic ''ought'' judgments.

In conclusion, I am perfectly content to call the epistemic concept I am interested in analyzing a normative concept—I have surely invited that label by distinguishing metaepistemology from normative epistemology. But it is important that we not make too much of this admission. In particular it is important that we not force the analogy between epistemology and ethics. Many of the most important ethical concepts, including the concept of what one ought to do, have at least some plausible analyses that make these concepts parasitic upon an understanding of epistemic concepts. Although the judgment that a belief is irrational or unjustified is often taken to be a criticism of the person or belief, there is clearly no conceptual connection between judging of a belief that it is irrational and blaming or criticizing the person who holds that belief. And even if we distinguish criticism of a belief from criticism of the believer, it is still by no means clear that there is a conceptual connection between judging a belief epistemically irrational or unjustified and criticizing the belief, except in the trivial sense that when a belief is unjustified or irrational it is a belief one epistemically ought not to have. This is trivial because the meaning of the adverb ''epistemically'' is still unanalyzed, leaving open the question of how strong the connection is between the epistemic ''ought'' and the ''ought'' of other paradigm normative judgments.

I have taken so much time discussing this issue for a number of reasons. First, as I indicated earlier, I am primarily interested in metaepistemological investigation into the concept of justified or rational belief and the implications of metaepistemological theories about such concepts for classical skeptical arguments against the possibility of having justified or rational belief. It is, however, important minimally to characterize as carefully as we can, without yet presupposing any particular metaepistemological view, the concept of justified or rational belief in which we are interested. And this requires that we carefully distinguish epistemic justification and rationality from other ways in which we can justify a belief or have reasons for holding a belief. By doing so we can at the start deflect the kind of criticisms that Plantinga raises against viewing the concept of justified belief as having a fundamental epistemic role.[21] Plantinga persuasively argues, for example, that the deontic concepts of obligation, duty, what one ought to believe or refrain from believing (in most ordinary uses of ''ought'') are not going to be the kinds of concepts with which one can plausibly analyze the third condition for knowledge. Furthermore, his discussion of the etymology of "justification" has convinced me that my and others' use of the term "justified belief" is potentially misleading precisely because it invites the kinds of confusion I have warned against in the preceding discussion. "Justifying" actions or beliefs sounds too much like "defending" actions or beliefs, where defending can get associated with defending from criticism or blame. As we shall see, understanding the senses in which epistemic justification is *not* normative is crucial in sorting through some of the pivotal debate between internalists and externalists.

For many of the reasons discussed earlier, the expression "rational belief" is probably preferable to the expression "justified belief" just because it avoids some of the latter's connotations.[22] But again, we must still be careful to distinguish different sorts of reasons we can have for believing something, including the distinctions discussed earlier between moral, prudential, and epistemic reasons for believing something. Notice, however, that when we discuss a problem like Pascal's wager, we have little difficulty explaining the nature of the ambiguities. People understand perfectly well that there are different kinds of reasons one can have for believing something, and that moral reasons and prudential reasons for believing in the existence of God are different from what I have been calling epistemic reasons. There is clearly a *prephilosophical* understanding of the distinction, even if explicating the distinction prior to our metaepistemological investigation is problematic.

Plantinga suggests introducing the technical expression "epistemic warrant" to distinguish it from justified belief and goes on to characterize warrant as whatever it is that must be added to true belief in order to get knowledge. I do think that a concept that I have been calling justified or rational belief is a constituent of some concepts of knowledge, but I am hesitant to define this as a necessary condition of epistemic justification or rationality for fear that I will beg what should at this stage be an open question, the question of whether knowledge is justified or rational true belief.

Is there any other relatively neutral characterization of epistemic justification or rationality that distinguishes it from other concepts of justification and rationality with which we are not primarily concerned? The most plausible suggestion takes us back to truth. Consider again Pascal's wager. The idea was that the cost of being a mistaken atheist is so high and the reward of being a correct theist so great that a prudent, reasonable person will take the precaution of believing in the existence of God. To make the argument a little stronger, let us suppose that somehow you know that if there is a God it is the kind of Judeo-Christian God who severely punishes disbelievers and greatly rewards believers and that God does not care much what caused you to believe in Him. Let us also introduce some moral reasons for believing in the existence of God. Let us suppose that you promised your dying mother that you would believe in the existence of God for as long as you lived and that promises are the sorts of things that create powerful moral reasons to behave in the way you promise to behave. As I said earlier, intuitively, we can see that even if we have strong prudential and moral reasons to believe in the existence of God, this has precious little to do with epistemology, with whether we are epistemically justified or rational in believing in the existence of God. How would we go about offering a preliminary explanation of the distinction?

Intuitively, the prudential and moral reasons are unrelated to epistemic reasons precisely because the justification or rationale they provide does not relate in the appropriate way to the *truth* of the proposition that there is a God. More specifically, epistemic justification and epistemic reasons for believing *P* must minimally *make probable* for the believer the truth of what is believed. The modal operator is important because we must admit that there are senses of "making probable" in which the prudential reasons we have for believing something increase the probability that what we believe is true. To take an example mentioned earlier, it may be that I have prudential reasons for believing that I will recover from an illness, reasons consisting in various desires I have concerning health and my epistemically justified belief that if I believe that I will recover the probability that I will recover will increase. Furthermore, there may be a statistical connection between my having these sorts of reasons for believing that I will get well and my getting well. Still this does not prevent us from understanding epistemic justification and rationality in terms of the relation of making probable, because it is necessarily the case that if *S* has some characteristic *X* which justifies or makes rational believing *P*, that characteristic *X* makes likely for *S* the truth of *P*.

In this preliminary discussion I was primarily interested in clearly identifying the concept of epistemic justification or rationality in which I am interested, and did not want to beg questions concerning correct metaepistemological positions. Both internalists and externalists have a reason to worry at this point. The emphasis on truth and probability might seem to prejudice the metaepistemological issue in favor of some version of externalism, perhaps some version of reliabilism. Crudely put, the idea behind reliabilism is precisely that there is some kind of necessary connection between the justifiedness of a belief and truth (although, as we shall see, the connection becomes more and more tenuous the more sophisticated the reliabilism becomes). Almost no one wants to hold the very strong view that a belief's being justified entails that it is true, but very crude reliabilists seem to want to hold something like the view that it is necessarily the case that most justified beliefs are true. It is necessarily the case, in other words, that justified beliefs are probably true. And is not that just a reformulation of our preliminary attempt to characterize epistemic justification or rationality?

This argument, however, proceeds too quickly. One of the critical underlying assumptions is the suggestion that justification rendering probable the truth of what is believed is to be explicated in terms of statistical probability of some kind. It seems obvious to me that there are concepts of probability that can be explicated in terms of frequency, but there is a controversy much older than the internalism/externalism controversy concerning the question of whether there are other species of probability more relevant to epistemology.[23] It may be that there is (and, as I argue later, internalists had better hope that there is) a concept of making probable that is not to be explicated in terms of frequencies. It may be, in other words, that there is some sense in which my justification for believing *P* might make probable the truth of *P* even if it is never the case that when I or anyone else has that kind of justification the proposition it justifies is true.

If internalists get nervous about the emphasis on probability, externalists who are paying close attention may not like the fact that I tried to sneak in a probability relativized to the believer. To be epistemically justified in believing *P*, *S*'s justification must make probable *for S* the truth of *P*. On at least one way of understanding "probability *for S*," the "for *S*" implies access of some kind by *S* to the probability. For now, let me simply acknowledge that the need for this relativizing of the probability is a matter of controversy that will be addressed in subsequent discussion.

In summary, the metaepistemological investigation I am interested in concerns the concept of justified or rational belief. I have acknowledged that the expression "rational" might be somewhat less misleading than the expression "justified," but I will continue to use the two interchangeably in this work, in part because by now there are a number of technical expressions using variants on the word "justified" that are an almost indispensable part of the literature on such controversies as foundationalism. *S*'s epistemic justification or reasons for believing *P* must make probable (for *S*) the truth of *P*. To describe someone as being epistemically justified or rational, unjustified or irrational, in believing *P* is not in and of itself to make *any* moral or prudential claim about what he ought to believe. It is not to praise or blame the person for having the belief. It is not, I think, even to praise or criticize the belief. Of course, given certain values, it may be extremely important to us to have justified as opposed to unjustified beliefs, just as given certain values, it may be extremely important for us to have smaller cars than bigger cars, but that makes neither "justified" and "unjustified" nor "bigger" and "smaller" value terms.

Philosophical Analysis

We are trying to accomplish two goals in this introductory discussion. The first is to describe in as neutral a way as possible the epistemic concepts with which we are primarily concerned in both our metaepistemological investigation and our discussion of skepticism. The other is to say at least something about philosophical analysis. Metaepistemology is an attempt to analyze the fundamental epistemic concepts of justified or rational belief. But what is involved in analyzing a concept?

If metaepistemology has a kind of priority over normative epistemology, metaphilosophy has that same kind of priority over metaepistemology. Metaphilosophy, however, is a Siren that can wreck one's philosophical work on the shoals of abstraction and indecision. It is no understatement to suggest that the discussion of what philosophers do when they analyze, explicate, and clarify deserves at least a book-length treatment. It is a subject I have addressed at some length elsewhere,[24] and so I shall be relatively brief in what I say here.

Philosophers who think of themselves as engaged in analysis talk about analyzing terms, sentences, concepts, properties, thoughts, propositions, states of affairs, and facts. The position one takes on the appropriate *object* of analysis in part determines how it would be most natural to formulate the metaepistemological questions. Thus a philosopher engaged in a metaepistemological investigation of epistemically rational belief might ask any of the following questions:

1. What is the meaning of "rational belief" in its epistemic sense?
2. What is the meaning of "*S* rationally believes *P*" when this is an epistemic evaluation?
3. What is the concept of epistemic rationality?
4. What is that property in virtue of which a belief is epistemically rational?
5. When I think that a belief is epistemically rational, what is it that I am thinking?

6. How shall we understand the proposition that a belief is epistemically rational?
7. What are the constituents of the state of affairs, *S*'s rationally believing *P*?
8. What are the constituents of the fact that makes it true that *S* rationally believes *P*?

Depending on how one formulates the metaepistemological question, one will have further metaphilosophical questions to which one must have implicit answers. If analysis consists in finding constituents (of a thought, concept, property, state of affairs, or fact), for example, one should say something about the relevant relation that the constituents must have to the whole that they constitute.

I have argued elsewhere that one must not take facts to be the primary objects of philosophical analysis. The reason, in short, is that analysis can proceed perfectly well even if we suppose there are no facts of the sort that would make true the assertions we are interested in understanding. The metaepistemological investigation into the nature of knowledge is perhaps a paradigm of analysis that often does not presuppose the truth of any knowledge claims. If it did, we would simply beg the question against all sorts of interesting skeptical arguments. But if we do not presuppose the truth of any knowledge claims in performing our analysis of knowledge, we cannot think that our primary concern is with the constituents of *facts* of the form '*S* knows that *P*.' At least we cannot if facts are what make true assertions true. And I reserve the term ''state of affairs'' to refer to those complexes that some philosophers call nonexistent facts, or (worse) false facts.

For phenomenological reasons, I have argued, it is difficult to construe philosophical analysis as simply an attempt to describe faithfully the constituents of *any* determinate ''object'' before the mind. It seems implausible to suppose that when one begins a philosophical analysis one brings before one's consciousness a property, concept, thought, proposition, or state of affairs which one then begins to dissect.[25] It is implausible for at least two reasons. First, one need only ask oneself when one begins one's search for causation, goodness, moral obligation, epistemic rationality, knowledge, or what have you, whether there pops into mind anything that is just there to be described. Phenomenologically, it seems to me that one often starts with nothing other than language before one's consciousness, with the barest idea of how exactly one is using that language. Second, any such metaphilosophical view of analysis has an extraordinarily difficult time resolving the paradox of analysis. The paradox, briefly stated, is this. How can a philosopher have so much difficulty finding the correct analysis of something X when by hypothesis to even formulate and understand the question "What is X?" one must already know what an X is? How can so many philosophers end up providing such radically different answers to a question like "What is causation?" if they all began with the same thing (a property, concept, thought, or state of affairs) before their consciousness?

I am not arguing that there are no such things as facts, states of affairs, propositions, thoughts, concepts, or properties. As you will discover, I accept a metaphysics that embraces many of these ontological commitments. I argue only that one has far too crude a conception of analysis if one thinks one can *begin* one's analysis by leaving the level of language. The view I defend requires that we take a "linguistic turn" with respect to understanding analysis, albeit a linguistic turn that recognizes a crucial role for nonlinguistic thought. Roughly, the idea is that in performing philosophical analysis we are performing egocentric meaning analysis. When I ask myself what epistemically rational belief is, I am seeking to discover the semantic rules that I follow when I describe a belief as epistemically rational. We dissolve the paradox of analysis by relying on the obvious distinction between following a rule and knowing what rule it is that one is following. Just as linguists go grey trying to figure out the syntactic rules *they themselves* follow when they use language, so philosophers can go grey trying to figure out the semantic rules they themselves follow when they use certain fundamental expressions.

In characterizing the rules one is seeking as semantic, it is crucial to stress that they are not rules that relate language to language. As I conceive of analysis, the philosopher's primary objective in providing a philosophical analysis is to relate the use of language to certain nonlinguistic states, to discover what I have called different-level meaning rules. The *ideal*[26] form of such rules is this: Regard "X" as a correct description of all and only those situations in which Y. To test whether one follows the rule, one must entertain nonlinguistic thoughts of various hypothetical situations in which the characteristics Y are present to see whether in fact in all the situations conceived of as Y one is disposed to regard "X" as a correct description of the situation.

In characterizing the philosophical enterprise as essentially egocentric, I mean only to emphasize that one is not *essentially* concerned with the way in which other people use the terms in which one is interested. I can engage in philosophical analysis even if I become a practicing skeptic with respect to the existence of other minds. It is, of course, a presupposition of almost all philosophical writing and debate that we are importantly alike in the different-level meaning rules we follow. I can engage in philosophical analysis if I am a skeptic with respect to the existence of other minds, but I might not talk quite so much with those bodies out there about those analyses. There *can*, of course, be mere verbal disputes between philosophers, and any account of analysis should accommodate the distinction between genuine philosophical disagreement and mere verbal dispute. The above account of analysis allows one to easily make that distinction. Verbal dispute occurs when some of the presuppositions of communication fail—when you and I presuppose that we are following the same different-level meaning rule, even though we are not.

This is far too brief a description of one conception of philosophical analysis, and I do not pretend to have defended it in any sort of detail. Again, I must refer the reader to my more detailed discussions of these issues.[27] Because I think metaphilosophical reflection is an important preliminary to philosophical analysis, I thought it important at this early stage to lay a few cards on the table. Before ending this discussion, I would also like to qualify and clarify some of my earlier positions.

First, it should go without saying that we must be sensitive to ambiguity in trying to figure out what we mean by epistemic rationality. And we should be equally sensitive to the distinctions among ambiguities. Some meanings are closely related to others. Some meanings are derivative from other meanings. It is important that in a metaepistemological investigation we not be driven to find *the* correct analysis of epistemic justification and rationality. It seems almost evident that there are a number of useful and interestingly different concepts of epistemic justification and rationality that find expression in both ordinary epistemological discourse and more technical philosophical discussion of epistemic problems. In this connection, it might be useful to mention a point that I emphasize later. It is that an interesting concept (use of an expression) can come into existence precisely because of the development of a field of inquiry. Just as in physics it often becomes useful to introduce a concept and even to start using a term in a way that it is not usually used,[28] so in *philosophy* we might start talking about justified or rational belief in a way different from the way in which these terms are ordinarily used. It seems clear that philosophers have interests *qua philosophers* that make it important to understand certain concepts in a way that makes them philosophically interesting.[29]

Second, there is no reason to suppose that the semantic rules one follows, even for fundamentally interesting philosophical concepts, are *simple*. In an excellent discussion of the implications of recent empirical discoveries in cognitive science for traditional conceptions of philosophical analysis,[30] Bill Ramsey suggests that the classic criteria for a successful conceptual analysis involve coming up with a small set of conditions (he says properties), usually a conjunction, that are individually necessary and jointly sufficient for the application of the concept we are interested in understanding, and that are immune to familiar sorts of "intuitive" counterexamples that rely on thought experiments. He goes on to argue that if the results suggested by some studies in cognitive science are correct, there may be reason to think that it is simply not possible to discover successful philosophical analyses with respect to many concepts. What are these results? They proceed from the observation that our judgments about whether or not something falls under a concept are rarely an all-or-nothing matter. Furthermore, the characteristics we take to be relevant to whether or not something is a paradigm X seem to vary from context to context.[31] There often does not seem to be any *necessary* condition for something falling under a certain concept. No matter what characteristic is missing, enough other characteristics can make up for it. While there is no agreement among cognitive psychologists on the theoretical model to explain these empirical findings, more and more psychologists (and philosophers) are turning to *prototype* theory.

There are all kinds of different models of how to understand prototypical representation. Ramsey refers to Smith and Medlin's (1981) "probabilistic" or "featured-based" account. On this account the prototype representation of an X consists in an "abstracted set of features," say, A, B, C, D, and E. Whether or not something is judged to be an X depends on how many of these characteristics the thing is thought to possess. But unlike traditional philosophical views that suppose that there are "neat" necessary and sufficient conditions for something being an X, this view recognizes that the features A through E can have different weights. Although it creates a misleading impression of precision, one could think of assigning various numbers to A through E to reflect the strength or importance the characteristic has in determining whether or not something is an X. The idea would be that one could compensate for the absence of even a very important characteristic, A, for example, if the thing possessed enough other characteristics whose weights added up to a weight equal to A. Furthermore, the weighting itself might be conditional. So A might have a very strong weight only if certain other characteristics, C and D, are present. In their absence it might be relatively insignificant. One can easily understand how on such a model one could make sense of things being better or worse examples of *X*. Two things might both meet the threshold for being an *X*, but the combined weight of the relevant characteristics of one might be greater than the combined weight of the relevant characteristics of the other. Prototype theory of concept formation can recognize more complicated relationships that exist between the relative importance of the characteristics that determine whether or not something is an *X*, but this example will suffice to make the observations I want to make.

It seems to me that something like this model of representation is extremely plausible for at least many concepts. It is one of the relatively few cases where an empirical investigation can shed light on a philosophical problem.[32] It is not clear, however, that the view really is inimicable to the search for necessary and sufficient conditions, or in the language I prefer, the search for the rule we follow having the form ''Regard '*X*' as the correct description of all and only those situations possessing characteristics *Y*.'' As Ramsey might admit,[33] the appropriate conclusion to draw might be only that the rules we follow in applying language to the world are extraordinarily complex. They may even be ''open-textured'' in something like the sense that the later Wittgenstein tried to explain. It may be, that is, that we can conceive of certain hypothetical situations in which we are disposed to say neither ''That is an *X*'' nor ''That is not an *X*,'' but rather, that we do not have the slightest idea whether or not we would call that an *X*. As philosophers became more and more inventive and imaginative in their Gettierstyle counterexamples to revised justified true belief accounts of knowledge, we can probably all think of appeals to prephilosophical intuitions about the correctness or incorrectness of a knowledge claim that left us shrugging our shoulders. Ramsey was primarily concerned with convincing us that prototype theory might rule out the possibility of presenting nice, neat conjunctive analyses of the relevant properties that determine whether or not something is an *X*, and in that he may have been successful. And like the syntactic rules we follow in forming sentences, rules that also can be so complex as to almost defy description, the semantic rules we follow in using language may defy a final complete description. But that need not stop us from sketching partial accounts and delineating the kinds of properties that seem relevant to the determination of whether or not a given concept applies. We can even do considerable work through thought experiments speculating about the weight a given characteristic has when it comes to determining the application of a term or concept.

Notes

1. Obviously, concepts can fall under concepts that are not constitutive of them. Moore was involved in no inconsistency when he held that the simple property of being good exemplified a number of other more abstract properties (e.g., being nonnatural).

2. See Armstrong 1973.

3. See, for example, Butchvarov 1970 and Plantinga 1992, p. 44.

4. See Sosa 1974 and 1991, Chisholm 1977, chap. 6, and Moser 1989.

5. See Hockett 1992 for a detailed discussed of this issue.

6. For an interesting discussion of the importance of appreciating ambiguities of perspective, see Foley 1987, pp. 123–24.

7. See Dretske 1970, pp. 1015–16.

8. Particularly in Plantinga 1992.

9. Ayer 1956, p. 31.

10. The history of this flirtation is long and complicated. It seems that Firth 1959 convinced Chisholm to abandon the suggestion, but even very recently Chisholm returns to ethical concepts by way of illuminating epistemic concepts. In the final analysis, I think Chisholm is prepared to live with his more common statement that fundamental epistemic concepts are indefinable. One can explicate one epistemic concept with other more fundamental concepts, but there is no way of breaking out of the circle of epistemic concepts. See Chisholm 1977, p. 12.

11. My colleague Laird Addis is probably an exception. Based on our conversations, I understand him to take the view that on *fundamental* normative issues of epistemology there is no truth or falsehood.

12. I use this more cautious term to allow for metaphilosophical positions concerning the nature of analysis that I will be addressing shortly.

13. Alston 1988 argues that one should not hold that one has epistemic obligations because one does not have the necessary control over belief. For an opposing view see Helm 1994. Helm admits that beliefs are seldom directly under our control, but he argues that they often flow from ''belief-policies'' that are freely chosen and for which we are responsible.

14. See Foley 1987.

15. Foley will do it in terms of the beliefs about efficacy that result from ideal reflection.

16. Foley 1987, p. 8.

17. For a detailed discussion of what constitutes actual consequence consequentialism and what differentiates it from other versions of consequentialism see Fumerton 1990.

18. A sense that is still distinct from our evaluation of the moral character of the agent.

19. For a detailed discussion of these important derivative concepts of morality and rationality, see Fumerton 1990 and Foley 1990.

20. For a useful critical discussion of possible conceptual connections between epistemic evaluation and judgments of obligation, praiseworthiness, and blameworthiness, see Alston 1988, Plantinga 1988, and Feldman 1988.

21. See, again, Plantinga 1992.

22. In the old sense of "connotations."

23. See Keynes 1921 for a clear statement of an alternative to a frequency conception of probability. There is an excellent discussion of concepts of probability (including an attempted refutation of Keynes) in Russell 1948. See also Pollock 1987, chap. 4. I return to these topics in some detail in chapter 7.

24. See Fumerton 1983.

25. This seems to be the way Moore thought of philosophical analysis throughout his career. See his famous discussion of how one begins the task of analyzing goodness in Moore 1903.

26. I will explain my reason for emphasizing "ideal" momentarily.

27. Fumerton 1983.

28. It is *possible*, for example, that the physicist really now means by "water" stuff with molecular structure H_2O. It seems clear to me that this couldn't have been what the physicist meant by "water" when the hypothesis that water is H_2O was *empirically* tested. (I think these alleged shifts in meaning occur much less frequently than philosophers suppose—I do not really think that physicists mean by "water" "H_2O" now, because I do not think that however much confidence they have in their hypothesis they regard it as unfalsifiable.

29. In *Unnatural Doubts* (1991), Williams wants to convince us that the skeptic's doubts are indeed unnatural and that they are expressed without the kind of context that gives "natural" claims to knowledge their (context-relative) meaning. But it is crucial to remember that philosophy *provides* a context in which one not only can but should worry about things one normally does not question. The philosophical context will make some concepts of knowledge and rational belief irrelevant to philosophical concerns.

30. Ramsey 1990.

31. Ramsey cites, among others, Rosch 1975; Rosch, Simson, and Miller 1976; Barsalou 1985; Rosch 1978; Rosch and Mervis 1975;, and Rosch and Shoben, 1983.

32. Although even here the data suggesting this account of concepts was readily available to philosophers. Indeed, the idea was foreshadowed in many respects by the work of the later Wittgenstein who emphasized the implausibility of holding that there are tidy necessary and sufficient conditions for the application of ordinary concepts.

33. I am basing this speculation in part on conversations I have had with him about the paper.

Chapter Two

The Structure of Skeptical Arguments and Its Metaepistemological Implications

Kinds of Skepticism

My primary concern in this chapter is to sketch what I take to be the most interesting form of skeptical argument and to examine in a preliminary way the metaepistemological presuppositions, if any, of skeptical arguments. I also attempt to address here some important challenges to the skeptic's method that do not explicitly rely on metaepistemological positions.

I hasten to emphasize that my concern in setting forth what I take to be the most interesting form of skeptical argument is not primarily historical. I do think that one can find the kind of skeptical argument I discuss developed in some detail by the Modern philosophers, and that its paradigm expression can be found in the writings of David Hume, but I will not defend this claim. Much of my view about the structure of skeptical argument also parallels the excellent discussion of this issue in chapter 2 of Ayer's classic book *The Problem of Knowledge*. I also remind the reader that the kind of skepticism in which I am primarily interested involves claims not about knowledge but about the epistemic rationality of belief. I have already observed that the historical discussion of skepticism is couched mostly in terms of what one can or cannot know. I am interested in what one can or cannot rationally believe.

Perhaps we should begin by making some familiar distinctions between kinds of skepticism. One distinction we have just marked—the distinction between skeptical claims about knowledge and skeptical claims about epistemically justified or rational belief. Let us refer to these two kinds of skepticism as *weak* and *strong* skepticism, respectively. As we noted earlier, the claim that one cannot *know* something (particularly if one adds the appropriate dramatic emphasis to "know") is relatively weak from the philosophical perspective in that one can easily concede this sort of skepticism with a kind of resigned respect for the fallibility of human consciousness. The claim that one cannot even rationally believe that something is true is much stronger, much more extreme. It seems to give the person who wants to continue believing what common sense requires nothing on which to fall back.

In addition to distinguishing strong and weak skepticism, we can also distinguish global and local skepticism. The global skeptic makes a claim about our epistemic access to *all* truth. Specifically, weak global skepticism maintains that one has no knowledge of anything. Strong global skepticism maintains that one has no epistemically rational beliefs about anything. Local skepticism is skepticism (weak or strong) with respect to a given class of propositions. Thus, we can be a skeptic with respect to propositions about the physical world, the past, other minds, the future, theoretical entities in physics, the existence of God, or any other subclass of propositions.

Skeptical claims can be made with or without modal operators. Thus the skeptic can claim only that we do not have knowledge or rational belief or that we *cannot* have knowledge or rational belief. When we examine more closely the kinds of arguments skeptics advance in support of their claims, we shall see that they typically support the stronger modal claims.

One must also distinguish first-level skepticism from second-level skepticism, or skepticism from what we might call metaskepticism.[1] Second-level skepticism involves skeptical claims about whether or not we have knowledge or rational belief. It is argued by some (even some externalists) that if certain versions of externalism are true, it may make first-level knowledge or rational beliefs *possible* only to invite skepticism about whether or not one ever has such knowledge or rational belief. Some might hope to concede the externalists' claims at the first level but allow for the legitimacy of traditional skeptical concerns at the next level. I attempt to show later on that from the philosophical perspective, nothing of any interest should change when one moves up a level and that the widespread feeling that something *does* change when one is operating from within an externalist framework has enormously significant implications.

So again, the kind of skepticism with which I am primarily concerned is strong skepticism. Historically, I think there have been very few *global* strong skeptics. It is, of course, the most paradoxical of skepticisms because it entails that one has no epistemic justification for believing it. Whether we should conclude that it is therefore of no philosophical interest is something that we discuss toward the end of this chapter when we examine various charges of self-refutation leveled against the skeptic. The vast majority of skeptics, I argue, have actually presupposed knowledge or justified belief with respect to some class of propositions. Skeptics in the empiricist tradition almost all seemed to presuppose unproblematic access to occurrent mental states. Indeed, the presupposition was so complete that one rarely even finds the Modern philosophers *raising* the question of whether or not one can know that one is in a certain subjective mental state. Furthermore, almost all skeptics seemed to presuppose knowledge of at least logical relations. They seemed to presuppose that one can recognize or ''see'' contradiction, at least some simple necessary truths, and at least some simple entailments.[2] As we shall see, the question of whether or not skeptics can ''contain'' their local skepticism is a matter of some controversy.

The Structure of Skeptical Arguments for Strong Local Skepticism

If one examines classic arguments for strong local skepticism, one can discover, I think, a recurring pattern. First the skeptic indicates the class of propositions under skeptical attack. Then the skeptic attempts to exhaustively characterize the most plausible candidate for something that could conceivably justify, or make rational, belief in this kind of proposition. Next the skeptic attempts to drive a logical wedge between the available justification and the proposition it is supposed to justify. The wedge is logical. The claim at this point is only that the justification available for that belief does not logically guarantee the truth of the proposition believed. It is conceivable that someone has precisely that sort of justification even though the belief in question is false. At this point, the Cartesian skeptic might end the argument with the weak skeptical conclusion that it is not possible to know *with certainty* the proposition believed. But this conclusion does not get one strong skepticism. The strong skeptic goes on to argue that the logical gap cannot be bridged using any legitimate nondeductive reasoning.

Let us try to illustrate the kind of skeptical argument discussed above with a few examples, and let us begin with familiar epistemological problems concerning our access to the external world. Our strong skeptic with respect to the physical world argues that it is not epistemically rational for us to believe any proposition asserting the existence of a physical object. To what evidence might we appeal in trying to justify our belief in the existence of some object? The best evidence we could possibly get (according to common sense) is the testimony of our senses. The presumption is that if I cannot rationally believe that there is a table in front of me now when I seem to see and feel a table, there is no proposition describing a physical object that it would be epistemically rational for me to believe. But does any number of truths about the phenomenological character of my subjective and fleeting sensations ever logically guarantee the truth of any proposition describing the physical world? The answer, the skeptic argues, is clearly no.

To support this answer the skeptic will often appeal to the famous *skeptical scenarios.* A skeptical scenario is simply a description of a perfectly intelligible hypothetical situation in which someone has the best possible justification for believing a proposition about the physical world, even though that proposition is false. In the case of beliefs about physical objects, the skeptical scenarios describe hypothetical situations in which one has the best possible evidence in the form of sensation that, for instance, the table exists, even though it does not. The great fascination of skeptical arguments no doubt is owed in large part to their exotic appeals to the possibilities of dreams, hallucinations, malevolent demons, brains in a vat, telepathic powers, and the like. No matter how vivid my visual and tactile sensations may seem to me right now, who could deny that it is at least conceivable that I have these qualitatively same sensations in a vivid dream or in a drug-induced hallucination? And if I am dreaming or hallucinating, it would be mere chance that the table I take to be there exists.

Mad neurophysiologists with futuristic knowledge of the workings of the brain provide particularly useful grist for the skeptic's mill. Many of the antiskeptics have an almost slavish devotion to the dictates of science, and science seems to tell us that it is brain events that are the immediate causes of (or, on some views, are identical to) sensations. By stimulating the relevant part of the brain in the appropriate way, it seems in principle possible to produce the very electrical discharge that will cause me to seem to see a table. If we "tickle" another region of the brain in just the right way, we can get a tactile "table" sensation. Indeed, if our neurophysiologist of the future has enough skill, sophisticated instruments, and knowledge of the brain, there is no reason to suppose that the brain could not be played like a piano to produce the extraordinarily complex set of sensations associated with visiting the Grand Canyon. And if it can be done, it can be done surreptitiously without the knowledge (or memory, if the knowledge once existed) of the subject. The intelligibility of the hypotheses seems hardly in question. It is the stuff of some extremely good and utterly *intelligible* literature and cinema.[3] To be sure, the skeptic's appeals to such possibilities are not considered unproblematic in the context of a skeptical argument, and we shall examine some of the complaints shortly. Still, it seems to be almost obvious that one can distinguish sensations from truths about the physical world and that no conjunction of truths about sensations will ever entail a truth about the physical world.

But so what? Epistemically justified or rational belief does not require the inconceivability of error, and even weaker concepts of knowledge defined in terms of justified true belief seem to make knowledge perfectly compatible with the conceivability of error. It is here that Hume took the skeptical concern to its natural conclusion. If sensations can occur in the absence of the physical objects we take them to indicate, what reason do we have for supposing that it is even likely that when we have certain sensations, certain physical objects exist? Well, how do we establish one thing as evidence for the existence of another? Perhaps the most familiar pattern of inference we employ to answer such questions is inductive argument. We take dark clouds to be a good indication of an imminent storm because in the past we have *observed* a correlation between the presence of such clouds and subsequent storms. We take the sound of barking to be a reliable (if not infallible) indicator of the presence of a dog because in the past we have *observed* a correlation between the occurrence of such sounds and the presence of a dog.

If we use this model to understand our reasons for relying on sensation as an indicator of physical objects, then to avoid strong skepticism we would need to make plausible the claim that we have in some sense *observed* a constant or near constant correlation between the occurrence of certain sensations and the existence of certain objects. But this, of course, we cannot do. We cannot step outside sensation to compare the sensation with the physical object it is supposed to represent. To use an inductive argument for the conclusion that sensation is a reliable indicator of the presence of physical objects, we would need access to physical objects that is independent of sensation. Without such access we could never discover the necessary constant conjunction of sensation and object. But we have no access to the physical world except through our sensations. Assuming we have unproblematic access to the past, we can perhaps correlate sensations. We can discover all sorts of interesting connections between visual, kinesthetic, tactile, auditory, gustatory, and olfactory sensations. But we can never step outside our mental states in order to correlate a mental state with something other than a mental state. And in the absence of our ability to discover correlations between the mental and the physical, we will never be able to rationally believe that there is a connection between the two.

In developing the above skeptical argument, one does need to establish the intelligibility of sensations occurring in the absence of the physical objects we take them to represent. Notice, however, that one can take the argument seriously and remain neutral on many of the controversies concerning the metaphysical analysis of sensation. One can think that the visual sensations that occur in the absence of physical objects involve our being related to another kind of object, a sense datum, or one can think that visual sensation should be understood as a nonrelational property of the mind or self (the so-called appearing or adverbial theory of sensation). All that one needs to take the skeptical challenge seriously is some understanding of sensation that allows us to speak meaningfully of the occurrence of sensations in the absence of physical objects.

Although epistemological problems of perception have occupied center stage in the history of skeptical challenges, epistemological problems concerning our access to the past through memory almost certainly have a more fundamental logical place in the ordering of skeptical issues to be resolved. The problem has received far less attention from philosophers partly because it is much less obvious how to characterize the nature of the available justification. If we suppose for a moment that there is such a thing as a memory "experience," an experience that can be veridical or nonveridical with respect to the past, the skeptical argument involving the past will closely parallel the argument for strong skepticism with respect to the physical world. *Ultimately*, in reaching conclusions about the past one must rely on what one seems to remember. But one scarcely needs an argument (particularly when one gets to be my age) that memory is fallible. One can have a *vivid* apparent recollection that one did something even if one did not. If an argument is needed, we can return to the apparent causal dependency of experience on brain events. If one can in principle produce nonveridical sensations by stimulating the brain, one can presumably find that region of the brain responsible for memory "experiences" and produce them at will. But if the occurrence of memory experiences is logically compatible with the events we seem to remember not having occurred, then what reason do we have for thinking that such memory experiences are reliable indicators of past events? It is tempting to rely again on inductive reasoning, but it would seem that such reasoning is once more unavailable. An inductive justification for the reliability of memory would proceed from a premise describing correlations between *past* memory experiences and the events we took them to correctly represent. But the skeptic wants to know what your reason is for supposing that in the past memory has typically been reliable. And, of course, the skeptic wants you to answer that question without begging the question, that is, without relying on memory. But if one cannot rely on the fact that one remembers having veridical memory experiences to justify one's belief that a memory experience is a reliable indicator of a past event, how could one ever get away from the present to gain access to the past?

Notice how much more fundamental the problem of memory is than other epistemological problems. Almost all candidates for resolving the problem of perception minimally presuppose that we have access at least to past sensations. Discussion of the notorious problem of induction again almost always presupposes that we have access to past correlations between properties in order to ask how one can justifiably project these past correlations into the future. In the context of worrying about how we can justify our belief in other minds relying only on observational knowledge of physical behavior, epistemologists typically "give" one knowledge of past correlations between one's own behavior and one's own mental states, that is to say, they presuppose that there is some solution to both the epistemological problems of perception and of memory. And philosophers of science who worry about the possibility of justifying belief in hypotheses that deal with microphenomena that are in principle unobservable (compared with the way in which ordinary macro-objects are presumed to be observable) typically assume the legitimacy of our conclusions concerning the macro-sized objects of the physical world, the past, and projectibility of observed correlations.

I suggested that the relative lack of concern with epistemological problems concerning memory might be due in part to the difficulty one has formulating the problem. As long as one supposes that there are such things as memory experiences, the skeptical argument goes relatively straightforwardly. But the existence of memory experience is far from uncontroversial. To be sure, philosophers do not agree much on how to understand visual, kinesthetic, tactile, gustatory, and olfactory sensations, or even the intelligibility of talk about the occurrence of sensations without physical objects. But that something new and in some sense *occurrent* comes into existence when I open my eyes, and ceases to exist when I close them, seems hardly problematic. It is far less obvious that when I remember putting my car keys on the desk something came into existence at a certain time, which can be meaningfully described as an occurrent mental state of seeming to remember having done something. As we shall see, the problematic nature of the presupposition that there are memory experiences can contribute to the attraction that externalist analyses of knowledge and justification have for so many philosophers.

Metaepistemological Presuppositions of the Skeptical Arguments

It seems to me that reflection on the above examples of skeptical reasoning strongly suggests that the skeptic relies implicitly on a principle that I call the principle of inferential justification:

> To be justified in believing one proposition *P* on the basis of another proposition *E*, one must be (1) justified in believing *E* and (2) justified in believing that *E* makes probable *P*.

Arguments for strong local skepticism typically invoke clause 2 of the principle first and then often counter proposed attempts to satisfy clause 2 by relying on clause 1. Thus the argument for strong skepticism with respect to the physical world relies on clause 2 by insisting that a belief in physical objects *inferred* from what we know about the character of our sensations is rational only if we have some reason to suppose that there is a connection between the occurrence of certain sensations and the existence of certain objects. When one attempts to inductively infer the existence of such a connection from a premise describing past correlations between sensations and objects, the skeptic invokes clause 1 of the principle to challenge our justification for believing the premise of that inductive argument. The argument for strong skepticism with respect to the past also relies on clause 2 insisting that any justified conclusion about the past inferred from what we seem to remember must include justification for believing that memory experiences are a reliable indicator of past events. Again, when an inductive justification of such reliability is attempted, clause 1 of the principle is invoked to challenge our ability to rationally believe the premise of the inductive argument, a premise that will describe past events and again require an inference based on memory.

It is tempting to think that the question of whether or not one accepts the principle of inferential justification determines whether or not one is an internalist or an externalist in epistemology. It is certainly true, as we shall see, that paradigm externalists reject at least clause 2 of the principle. It is *not* true, however, that paradigm externalists reject clause 1. Nevertheless, the above skeptical arguments do not get off the ground unless it is presumed that a reasonable conclusion based on premises requires a reasonable belief in the proposition that the premises make probable the conclusion. And on one reading, externalists avoid traditional skeptical problems by refusing to accept clause 2 of the principle of inferential justification.

The issue is, however, complicated. First, it is not clear that the principle of inferential justification constitutes a metaepistemological principle. On the face of it, one could accept the principle as a very general normative principle of epistemology. Second, it is not clear that one should simply define the internalism/externalism controversy in terms of whether one does or does not accept both clauses of the principle of inferential justification. In the next chapter I examine more carefully the controversy between internalists and externalists in epistemology. I try to define precisely the controversy, or more accurately a number of controversies, and we return to the question of which version of internalism, if any, classic skepticism presupposes. In chapters 3 through 7 we attempt to arrive at some conclusions about the philosophical plausibility of various metaepistemological views and the position they leave us in with respect to answering the skeptic. But before we leave this preliminary discussion of skeptical arguments, I want to emphasize and clarify certain features of the arguments. I also want to reply to a number of objections that are leveled against skepticism, objections that do not explicitly focus on the principle of inferential justification and any metaepistemological implications that acceptance of that principle might be thought to have. My goal is to leave the skeptic in as strong a position as possible when we consider the question of whether the externalist revolution in epistemology is the only effective way to circumvent the skeptical challenge.

Clarification of the Skeptical Arguments and Charges of Self-Refutation

Inference and Inferential Justification

If the skeptic does rely on the principle of *inferential* justification in presenting skeptical arguments with respect to our epistemic access to the physical world, the past, other minds, the future, and so on, he is obviously committed to the conclusion that the only justification available in all of these cases is inferential in character. This is a feature of the classical skeptical argument that has understandably come under considerable attack. In what sense do I actually *infer* the existence of the familiar objects around me now? Is it not even more strained to talk as if my belief about what I had for breakfast this morning involved some inference from propositions describing the phenomenological character of my present memory states? Many philosophers would even suggest that our beliefs about the conscious states of those around us are misleadingly described as involving inference.[4] When someone is writhing on the ground before me, I ''see'' the suffering directly. I hardly notice first a pattern of behavior, think about correlations between behavior and pain, and then reach the conclusion that the person before me is in pain. Of course, if we are relying on phenomenological evidence in order to determine whether it is plausible to maintain that an inference has taken place, it is equally problematic to suppose even that beliefs about the future typically involve inference. When I drink the water I expect it to quench my thirst. This expectation is probably (although not necessarily) *caused* by past associations of drinking water and subsequent diminution of thirst, but I can confidently assert that I never recall having attempted to list the occasions on which drinking water quenched my thirst in order to generate the premises of an inductive argument. My memory is so bad, I would be hard pressed to come up with more than a dozen such occasions, hardly enough to get me the kind of impressive correlation one should have for a strong inductive argument.

Whether or not the justification available for a given belief involves inference will be a difficult question to answer and will inevitably involve complex philosophical controversies. In characterizing a kind of justification as inferential, we seem to be implicitly contrasting that justification with some other kind of justification—noninferential justification. Obviously one needs an analysis of the distinction, and that we attempt to give in chapter 3. For now, let us make some relatively innocuous comments about ways of thinking about the distinction that bear on the reader's rather natural concern that the skeptic is talking about inference where there appears to be no inference.

The first distinction one might make when facing this sort of objection to talk about inference is the familiar distinction in philosophy of science between the context of discovery and the context of justification. It is not obvious that in order for one's justification to be inferential one must actually go through some process of conscious inference. It is one thing to arrive at a conclusion. It is another to justify that conclusion. It is almost certainly the case, it seems to me, that we do *not* usually formulate, and perhaps *never* have formulated, premises describing the phenomenological character of sensation as part of an attempt to gather evidence in support of our beliefs about the world around us. Indeed, it is probably seriously misleading to talk about my *belief* that there is a table before me and a wall behind me now. Philosophers have a tendency to oversimplify the range of intentional states that the conscious human mind can exemplify, and our language may even be inadequate to capture the subtle differences between them. There does seem to me to be a difference between belief and what one might call expectation. I do not so much believe that the wall is behind me as I expect it to be there. The expectation may consist in nothing more than the disposition to believe the proposition were I to entertain it, and the disposition to be extremely surprised if I were to turn around and not have the familiar range of experiences associated with there being a wall there. In any event, these beliefs (occurrent or dispositional) and expectations are still the kinds of things that can be justified or unjustified and we can still raise questions about the justification *available* to support such beliefs.

But surely, one might object, this is a little too slick. The skeptics and their opponents are raising questions about whether our *actual* beliefs are *actually* rational or not. And if we accept the framework within which the skeptics are asking this question, it would seem that their opponents run the risk of getting caught in the presupposition that if we are actually justified in thinking that there is table before us, we must have actually gone through some process of inference. The *availability* of a legitimate inference that we might have engaged in seems neither here nor there when it comes to the question of whether a belief *is* justified. This is perhaps a point well taken, and I argue later that a complete answer to it requires that we distinguish a number of different senses in which we can talk about a belief's being inferentially justified. In particular, it will be useful to distinguish an *ideal* sort of justification from derivative concepts of justification. For now, let us be content with these observations. First, nearly everyone who talks about inferential justification wants to allow that the beliefs that are involved in inferentially justifying another belief might be merely dispositional. Second, engaging in conscious consideration of some set of premises on the way to reaching a conclusion that is consciously thought of as following from those premises is not a necessary condition for the justification supporting that belief to be inferential.

The Role of Skeptical Scenarios

Although it is perhaps obvious already from the earlier discussion of the form of classical skeptical arguments, I want to emphasize the limited role that the familiar skeptical scenarios play in generating the skeptical conclusion.[5] As I presented those skeptical arguments, appeal to the possibility of dreams, hallucinations, artificially stimulated brains, and the like is primarily designed to support the conclusion that one has no *direct*, unproblematic access to the truth of the propositions under skeptical attack, and one cannot *deduce* the propositions under skeptical attack from the available evidence. The skeptic I am interested in is *not* presenting the following superficially similar argument for skepticism concerning the external world: The hypothesis that there is no table before you now but you are dreaming (hallucinating, having your brain artificially stimulated by a neurophysiologist, being deceived by an evil demon) contradicts your commonsense belief, and you do not know or have reason to believe that this alternative hypothesis is false. Therefore, you do not know or have reason to believe that you are really seeing a table. This argument simply invites the kind of response that Peter Klein developed so plausibly in his excellent book *Certainty*. That disarmingly straightforward response is to announce that we do have epistemic reason to reject the skeptical scenarios, a reason that consists in our being justified in accepting commonsense conclusions about the physical world. One might just as well argue, after all, that since the skeptical hypotheses are incompatible with the dictates of common sense and the dictates of common sense are epistemically rational, then the skeptical hypotheses are irrational. We can deduce their falsehood from the commonsense premises rationally believed. There is nothing wrong with Klein's *strategy* for defeating this form of skeptical argument, provided that he can establish the crucial conclusion that our commonplace beliefs are epistemically rational.

Again, however, as I construe the skeptic's appeal to the intelligibility of skeptical hypotheses, they are designed only to show something about the nature of our justification for believing propositions about the physical world, the past, other minds, and so on, namely that such justification involves nondeductive inference. If this conclusion can be reached, then one forgets about skeptical scenarios and invokes straightforwardly the principle of inferential justification. Once we have agreed, for example, that the occurrence of our sensations is perfectly compatible with there being no physical world, the skeptic can invoke the principle of inferential justification in order to request some positive reason to suppose that there is at least a *probabilistic* connection between the occurrence of certain sensations and the existence of certain objects.

Notice that here there is no burden of proof. Many discussions of skeptical challenges begin with a frantic jockeying for position. The antiskeptic wants the skeptic to give some positive reason for supposing that there is no physical world, and in the absence of such argument proposes that we continue with the beliefs that we are in any event disposed to have. The skeptic, however, armed with the principle of inferential justification, can adopt what seems to me the correct philosophical attitude that the principle of inferential justification plays no favorites, recognizes no special burdens of proof. The astrologer and the astronomer, the gypsy fortune-teller and the economic forecaster, the druid examining entrails and the physicist looking at tracks in cloud chambers are all expected to have reason to believe that their respective evidence makes probable their conclusions if the conclusions are to be rational. And you are expected to have reason to think that your sensations make probable the existence of the objects you take them to indicate if you are to be justified in believing the dictates of common sense.

Before leaving this preliminary discussion of the role played by appeals to skeptical scenarios, I should add that there is at least one attempt to satisfy the second condition of the principle of inferential justification that can give rise to a second purpose served by appeals to the intelligibility of such hypotheses. Whereas earlier in this century philosophers tended to emphasize the prominence of enumerative induction as the most obvious candidate for legitimate nondeductive reasoning, contemporary philosophers who realize the limitations of inductive reasoning as a means of regaining commonsense beliefs about the world have often turned to so-called reasoning to the best explanation.[6] The physical world, the past, other minds, lawful regularities, and theoretical entities are posited, the argument goes, as the best explanation for the order in which sensations come and go, the existence of memory experiences, the behavior of other bodies, observed regularity, and phenomena in the macroworld. In response to this gambit the skeptic will often request criteria for ''best explanation'' and turn again to skeptical scenarios to argue that there are always alternative explanations that compete with our ''commonsense'' hypotheses and that satisfy equally well the criteria of good explanation.[7] In chapter 7 we will examine in detail the appeal to best explanation as a solution to the skeptical challenge and discuss the way in which the skeptic might invoke skeptical scenarios to counter the use of reasoning to the best explanation.

Epistemological Commonsensism

I have suggested that the skeptic who invokes the principle of inferential justification wants a level playing field. In particular, the skeptic will not give the mere fact that you are inclined to believe a hypothesis any particular weight. There are, however, many philosophers who would argue that one must simply rule out skeptical conclusions from the start. The most common form of argument in analytic philosophy is the reductio. One objects to a philosophical position by pointing out that it has absurd consequences and is therefore absurd. But strong skepticism with respect to commonsense beliefs is itself patently absurd, the argument continues, and is a sufficient reason to reject any view, including a metaepistemological view, that leads to it. We might call the view that rules out skepticism from the start and evaluates metaepistemological views in part by the way in which they allow one to avoid skepticism, epistemological commonsensism. I return later to the suggestion that skeptical conclusions are absurd, but for now I want to make at least a preliminary comment on one sort of argument for it. The most obvious question the skeptic will ask is *why* we should assume at the outset that the beliefs we take to be justified are justified. The answer that we must start somewhere will no doubt not please a skeptic who is disinclined to start a careful reexamination of all of our beliefs with the presupposition that most of those we take to be justified are justified.

A somewhat more sophisticated answer involves appeal to science and evolution. We can assume, the argument goes, that most of our beliefs are justified because it is obviously evolutionarily advantageous to have justified beliefs and science tells us that what is evolutionarily advantageous has a high probability of occurring, other things being equal. Such an argument will not impress traditional skeptics, of course, because they will correctly point out that the pronouncements of science can be used to refute skepticism only if they are themselves justified. Their justification, however, presupposes solutions to the various problems the skeptic presents.

Whether or not science can refute traditional skepticism will, as we shall see, depend itself on the plausibility of certain metaepistemological views.[8] But for now I also want to remind the reader that even if we give ourselves full access to the pronouncements of science, it is not clear that science does tell us that it is evolutionarily advantageous to have justified or rational beliefs. Many of the empirical conclusions of science seem to suggest that much of what we expect or take for granted is "programmed" into us through evolution. If children had to reason deductively and nondeductively to the various conclusions they take for granted, their chances of survival would no doubt be rather slim. There is no reason to believe that we are not simply programmed to respond to certain stimuli with certain intentional states, just as lower life forms appear to be programmed to respond to certain stimuli with appropriate behavior. Now *given* certain metaepistemological views that we will discuss later, the causal origin of these spontaneous unreflective beliefs might be sufficient to make them justified, but one might also conclude that nature has simply no need to satisfy the *philosopher's* desire for having fully justified belief. One might, in other words, argue that if what science tells us is true, one might well expect that nature has probably not constructed us to believe only that which we have good reason to believe. Commonsense science might well tell us that through evolution nature has decided that it would be better for us to have true beliefs than justified beliefs.[9]

Of course, if we could *know* this we would again have reason to believe that most of our beliefs are true. My only concern here is to point out that our scientific *beliefs* are perfectly compatible with the conclusion that those beliefs are quite unjustified, and may well even suggest that conclusion, given certain metaepistemological views about what is required for justification. It is probably just this thought that led Hume to observe, with respect to the question of whether or not man should believe in a physical world, "Nature has not left this to his choice, and doubtless esteem'd it an affair of too great importance to be trusted to our uncertain reasonings and speculations."[10]

Charges of Self-Refutation

Epistemological commonsensism maintains that skepticism runs afoul of methodological constraints on epistemological investigation. The charge of self-refutation is a much more specific criticism of skepticism and must be treated very carefully. Let us begin by distinguishing two ways in which an argument might be charged with self-refutation. The first charge of self-refutation consists in the claim that the conclusion of a skeptical argument is inconsistent with the premises that are used to reach that conclusion, or in the claim that the very intelligibility of the skeptical conclusion requires that it be rejected. Let us say that an argument with this feature is formally self-refuting. If the conclusion of an argument really is inconsistent with its premises, then of course the argument either is formally invalid or has necessarily false premises. In either case an argument that is formally self-refuting in this way is always unsound. A more interesting and probably more commmon criticism of skeptical arguments is that they are *epistemically* self-refuting. Let us say that a skeptical argument is epistemically self-refuting if the truth of its conclusion implies that one has no justification for accepting its premises. But if one has no justification for accepting its premises, then the very principle of inferential justification upon which the skeptic relies implies that one cannot be justified in believing the conclusion by inferring it from the premises.

Of the two ways in which one might assert that skepticism is self-refuting, the charge of formal self-refutation is going to be the hardest to make stick. The most common allegations of formal self-refutation today involve claims about language and intentionality. Largely under the influence of the later Wittgenstein, a number of philosophers have explicitly or implicitly adopted what one might call a *contrast* theory of meaning. Roughly, the idea seems to be that a predicate expression ''*X*'' only has meaning if there are things that are both correctly and incorrectly described as being *X*. Thus, on my reading of Wittgenstein's private language argument, the fundamental objection to a private language has nothing much to do with memory. The problem is that a private linguist is the sole arbiter of how similar something must be to a paradigm member of a class to count as similar enough to be described in the same way. But as the sole judge it will not be possible to make a mistake, and where there is no possibility of error there is no possibility of getting it right. It is only meaningful to talk about the correct application of a rule if it can be contrasted with an incorrect application of the rule. If one applies this principle to dreams, hallucinations, and more generally, nonveridical experience, it will make sense to speak of nonveridical experience only if we are contrasting such experience with something else.

The contrast theory of meaning as stated is far too crude to evaluate. Suffice it to say, for our purposes, any remotely plausible version of the view would have to stress the modal operators. It will make sense to talk about dreams only if it is *possible* to have an experience that is not part of a dream. It will make sense to say of something that it is not a unicorn only if it is *possible* for something to be a unicorn. But to entertain the possibility that one is always dreaming, always hallucinating, or always being deceived by the senses is not to entertain the hypothesis that veridical experience is in any way *impossible*.

In response, the antiskeptic pushing this argument might appeal to a still more controversial verificationist theory of meaning. Roughly, the idea is that for X to be meaningful, not only must it be possible for both something to be X and something not to be X, but we must have *criteria* for distinguishing the Xs from the not-Xs. But for the more vague expression "criteria," this principle is, of course, familiar as at least a relative of the logical positivists' old verifiability criterion of empirical meaningfulness. On the face of it, however, the verifiability criterion of meaning, even in its very weak forms, has little to recommend it. As others have pointed out, we can entertain perfectly meaningful hypotheses that we could not in principle verify or disconfirm. Consider, for example, the proposition that there are things of which no one has ever or will have ever thought. What are our criteria for picking out the things about which we have never thought? Far from being meaningless, the hypothesis is entirely plausible. But if we can entertain and even believe such a hypothesis, why should we be unable to suppose that the causes of our sensations are radically different from any of which we have thought, and different in a way that would make *all* of our ordinary beliefs about the causes of sensations false?

Without focusing as much on language alone as a means of representation, Putnam also seriously questions the intelligibility of the skeptical conclusion given its premises. In the now-famous discussion of "Brains in a Vat,"[11] Putnam argues that if we were brains in a vat we couldn't assert that we are brains in a vat. It will not be possible to do justice to the argument here because it rests on a highly sophisticated, controversial, and incomplete theory of what is involved in one thing representing another. Given the attention the argument has received, however, I should at least indicate how I would respond.

Crudely, Putnam's idea is that for a mental state, a thought, to be a thought of X, X must causally interact with the state that represents it. Like most theories of representation, the principle would be qualified to allow for complex thoughts of nonexistent things that are "constructed" by the mind out of simpler thoughts. Hume, for example, thought that all simple ideas had their source in experience but allowed that one could form the idea of a unicorn without seeing one by "putting together" the idea of a horse and a horn. So, too, Putnam could allow that one might form the idea of things that do not exist by putting together the ideas of things with which one has had causal interaction. Within the framework of such a view, for my talk or thought of physical objects to mean what it does, my language and mental states must have interacted in the appropriate way with the physical world. Brains in a vat, by hypothesis, do not interact with the physical world in the way that would be relevant to allowing such brains to represent physical objects. But if brains in a vat cannot represent physical objects, then they cannot coherently frame the hypothesis (as we understand it) that they might be *brains* in a *vat*. By the same reasoning, they cannot coherently frame the hypothesis (as we understand it) that there might be no *physical world*. For these hypotheses to be genuinely meaningful, they would have to be false.

I have argued elsewhere that the theories of representation on which the argument rests are false, and for that reason, and because an excursion into the metaphysics of intentionality would take us too far afield, perhaps we should content ourselves with the observation that even if something like Putnam's conception of representation were correct, he would not get much ammunition for use against the skeptic. By now I think that almost everyone agrees that Putnam's original argument involved rhetorical ''stretch.'' For obvious reasons, even if we assume as true everything Putnam claimed about representation, it would still be possible for *me* to be a brain in a vat now, for *me* to have always been a brain in a vat, for *all* humans to be brains in a vat now, and for *all* humans to always have been brains in a vat. The reason all of these hypotheses are intelligible on Putnam's view is that our ability to represent can ''piggyback'' on prior representation. The relevant causal chains that allow us to represent an *X* can be extraordinarily convoluted. And the skeptic can surely get what is needed (particularly given the limited role skeptical scenarios play on my construal of the skeptical argument) from the intelligibility of the above skeptical scenarios.

The other observation I would make is that the ability to construct complex ideas of things that do not exist out of simpler ideas representing existents leaves enormous room for the skeptic to argue that hypotheses about the physical world might all be intelligible while at the same time being false. Put briefly, why should we not suppose that the concept of a physical object just is one of those innumerably many concepts of something that does not exist? On one historically prominent conception of the physical world, physical objects are *thought* of as the causes of certain sensations that stand in certain isomorphic relations with the sense data they cause. If one can construct the thought of phlogiston only to find out that there is nothing corresponding to it, why can one not construct the thought of a physical world only to realize that one has no reason to believe that there is such a thing? The ideas of sensation and causation might be traced to actual phenomena. But a complex idea formed out of these might not. I am not arguing for such an analysis of the concept of a physical object. I am only pointing out that in the absence of considerable argument to the contrary, there is every reason to suppose that if one can have ideas of nonexistent things, one can have the idea of a nonexistent physical world.

A more whimsical argument still for the incoherence of skeptical challenges to common beliefs is suggested by Davidson's much discussed remarks concerning interpretation. Davidson (1981) argues, perhaps not too implausibly, that in interpreting another's language I must (methodologically) assume that most of what that person asserts and believes is true. He then argues that even an *omniscient* being would need to employ such a principle of charity when interpreting ordinary claims about the world. But if an *omnisicent* being would need to believe that most of what we believe is true, then most of what we believe is true! The argument seems too good to be true and, of course, it is. Foley and I have pointed out that Davidson's argument actually needs a premise asserting the existence of an omniscient being to secure its conclusion.[12] To fully understand how so many people could have taken seriously the argument, we would need to examine more fully the presuppositions of the counterfactual conditional about the omniscient being. As this would take us too far afield, I simply refer the reader to our article on Davidson's "theism."

It was obviously going to be an uphill battle to make good the claim that there is something formally self-refuting in the skeptic's position. It would be more than a little odd to be able to reach the conclusion that a position with which people have been fascinated for thousands of years was in some deep sense literally unintelligible. The charge of epistemic self-refutation, however, is likely to be more significant. A skeptic who is in the business of undermining presuppositions about the rationality of our beliefs must be careful that the foundations on which the skepticism gets built are not undermined in the process. In what ways might an argument be epistemically self-refuting?

Consider again classical skeptical arguments for strong local skepticism concerning propositions describing the physical world. You will recall that a crucial step in the skeptical argument is to establish the possibility of sensations occurring without the physical objects we take those sensations to represent. In arguing that this is a genuine possibility, the skeptic appeals to the possibility of dreams, hallucinations, or surreptitious manipulation of that part of the brain directly responsible for producing sensations. But if the skeptic appeals to facts about the causal dependency of sensations on brain events, or facts about the hallucinogenic character of certain drugs, or previous occasions on which dream experiences seemed indistinguishable from the experiences of waking life, then surely the skeptic must be justified in believing these truths about the physical world in order to generate the conclusion that one cannot be justified in believing any proposition about the physical world. The skeptical conclusion, however, entails that the skeptic has no justification for believing these commonplace truisms about the causal conditions surrounding sensation.

Is there any way for the skeptic to avoid the charge of epistemic self-refutation? I think there is, and that it involves careful use of modal operators. To be sure, many of the classic skeptics (or philosophers seriously advancing for consideration skeptical arguments) appeared to appeal to facts about the physical world in advancing their argument. Thus, in his refutation of epistemological direct realism, Hume certainly did start talking about what happens to your visual sensation when you press against the side of your eye.[13] And Descartes does sometimes describe actual dreams he remembers having.[14] But given the purpose that appeal to these skeptical hypotheses is designed to serve, these philosophers need not have made any claims about what actually happens. One need only appeal to the *intelligibility* of vivid dreams, hallucinations, or the causal dependency of sensations on a physical world to establish that there is no logical connection between the occurrence of sensations and the existence of physical objects. Claims about possibilities are not in any obvious sense contingent, and would not fall within the scope of the propositions under attack by the skeptic advancing local skepticism with respect to the physical world. It must be conceded that the most convincing proof of a possibility is an actuality. Descartes's dream argument would hardly have fascinated us so much were we never to have vivid dreams. Indeed, Descartes probably would not have thought of the argument in the first place (nor would he even have been understood) were people not convinced that dreams do occur. And if people did not believe the scientific data on the brain, one would not get very far appealing to the possibility of producing sensation by manipulating the brain. But a clever enough philosopher could have formulated these hypotheses as possibilities even without the empirical evidence that the possibilities are actual and could have reached the appropriate conclusions about whether or not our access to the physical world was direct. Since the appeal to possibilities presupposes no knowledge or rational belief about the physical world, a skeptical conclusion based in part on such appeals involves no epistemic self-refutation.

Earlier, I argued that the problem of skepticism with respect to the physical world is not the most fundamental of the local skeptical challenges. Without rational beliefs about the past, one could not even learn anything about connections between sensations that one might hope to use in justifying belief about the physical world. But strong skeptical conclusions about the rationality of belief in past events is much more likely to end up involving epistemic self-refutation. The problem is more severe, of course, the stricter one's conception of what the past is. In the most extreme view, "now" refers to an instant in time, and the epistemological problem of access to the past is the problem of how one can reasonably infer the occurrence of *any* past events from data that is available to consciousness *now*, this instant. Reasoning itself, including skeptical reasoning, takes place over time. The skeptic moves step by step to a skeptical conclusion. If the skeptic implicitly accepts the principle of inferential justification, then to be justified in reaching a skeptical conclusion each step in the reasoning process must itself be justified. But if one has no justification for believing anything about the past, if one is an epistemic prisoner of present consciousness, how could the skeptic be justified in believing the premises of the skeptical argument?

One attempt to get around the problem is to expand "now" to include what is sometimes called a specious present. Consciousness is capable of grasping directly and immediately an expanse of time and a sequence of events that are included in that limited time. If the specious present is "large" enough and the skeptic's reasoning is quick enough, one might try to avoid the charge of epistemic self-refutation this way. Needless to say, it would not be easy to decide on the duration of a specious present. Unless it grows to an implausible size, the skeptic must think a lot faster than I can to escape epistemic self-refutation in this way.

Curley (1978) has argued that a skeptic can avoid charges of epistemic self-refutation by holding the premises of the skeptical argument to much lower epistemic standards than those challenged by the conclusion. This escape certainly would be in principle available to the philosopher arguing only for weak skepticism with respect to the past. One could easily reconcile the conclusion that we cannot know with absolute certainty anything about the past with the claims that one has some reason to believe some propositions about the past, and that those somewhat rational beliefs are sufficient to entitle one to accept rationally the skeptical conclusion. But this reply is clearly unavailable to the philosopher arguing for strong skepticism.

Ultimately, it may be that the charge of epistemic self-refutation will stick in the case of the most extreme skepticism about the past. The reason again is straightforward. As we already have had occasion to note, all strong global skepticism is epistemically self-refuting. If one concludes that one has no epistemic reason for believing anything at all, then it follows that one has no epistemic reason for believing that one has no epistemic reason for believing anything at all. Further, one has no epistemic reason for believing anything upon which one bases one's conclusion that one has no epistemic reason for believing anything at all. That is precisely why so few skeptics have been strong global skeptics. In particular, skeptics have almost always presupposed a kind of unproblematic access to some foundational empirical data and to the legitimacy of the reasoning on which their skeptical conclusions depend. But an extreme, strong skepticism with respect to the past challenges the unproblematic access to reasoning and seems to run the danger of spilling over into areas not under skeptical attack.

If the skeptic has difficulty denying the charge of epistemic self-refutation to this very fundamental sort of strong local skepticism, perhaps the next best move is to absorb it in a way that leaves the force of the skeptical argument intact. Is one in a position, after all, to dismiss an argument on the grounds that it is epistemically self-refuting? Well, if strong skepticism with respect to the past is epistemically self-refuting, then by definition, the skeptic is not epistemically rational in believing at least some of the premises of the argument. But if the argument is a valid argument and if the anti-skeptic *believes* the premises of the argument, the anti-skeptic is hardly out of the woods.

When I was a child I owned a ''magic'' eight ball whose function was to predict the future. You asked the eight ball a question that could be answered ''Yes'' or ''No,'' shook the ball, and a ''Yes'' or ''No'' floated to a transparent opening in the ball. Now suppose we lived in a culture in which people took the eight ball to be a reliable guide to the future. If you asked the eight ball ''Will it rain tomorrow?'' and the eight ball answered ''Yes,'' then, according to the members of our culture, you would be epistemically justified in believing that it will rain. Let us also suppose that our culture contains a few annoying skeptics who do not see what possible grounds one could have for supposing that the eight ball's answers are reliable predictors of the future, and who advocate strong skepticism toward conclusions reached via eight-ball reasoning. Finally, let us suppose that one day a skeptic gets the bright idea of asking the eight ball whether or not conclusions reached via eight-ball reasoning are rational, and the eight ball answers ''No.'' To the chagrin of the supporters of commonsense eight-ball reasoning, the results of the experiment are duplicated again and again. In what position are eight-ball reasoners left?

I suppose die-hard proponents of eight-ball reasoning can argue that our skeptic who concludes that it is irrational to believe the dictates of the eight ball, is certainly in no position to use the eight ball's "answers" to reach that conclusion. This seems right, of course. The skeptical argument that proceeds from observations about what the eight ball indicates and reaches a conclusion about the illegitimacy of eight-ball reasoning is epistemically self-refuting. But should the eight-ball reasoners be celebrating? Can they go on as before, trusting the predictions of the eight ball? If it is obvious that the skeptic's eight-ball argument is epistemically self-refuting, it seems equally obvious that the anti-skeptic cannot continue to embrace the unproblematic legitimacy of eight-ball reasoning. The skeptical argument has revealed an internal problem for the anti-skeptic despite the fact that the argument is epistemically self-refuting. In the same way, I would suggest, a skeptical argument with the conclusion that beliefs about the past are epistemically irrational is not something one can dismiss just because one concludes that even the skeptic would have to rely on some beliefs about the past in order to reach the skeptical conclusion. As long as the anti-skeptic shares belief in the premises that the skeptic acknowledges one has no epistemic reason to accept, and as long as the anti-skeptic has no reason to reject the legitimacy of the skeptic's reasoning, the anti-skeptic cannot simply dismiss the import of the argument on the grounds that it is epistemically self-refuting.

It is perhaps worth emphasizing again that as I presented them, most skeptical arguments for strong local skepticism are not epistemically self-refuting. The premises that the skeptic would need to believe rationally in order to infer the skeptical conclusion fall outside the class of propositions under skeptical attack. A fundamental and extreme sort of local skepticism concerning belief about the past may encounter difficulties with epistemic self-refutation, but the argument will still be a thorn in the side of the anti-skeptic until the anti-skeptic can figure out which of the skeptic's premises should be abandoned.

In the preceding discussion we distinguished charges of formal and epistemic self-refutation that might be leveled against the skeptic's arguments. In addition to these relatively precise criticisms of skepticism there is a somewhat more nebulous charge that is related to criticisms associated both with epistemic commonsensism and charges of self-refutation. That criticism attempts to denigrate the importance of the skeptical challenge by observing that it is in some way impossible to take skepticism seriously. In fact, most philosophers have never treated skepticism as a viable option. Even those who are interested in the skeptical challenge are primarily interested in finding the correct way of refuting what is assumed to be an illegitimate conclusion. Everything the skeptic *does* belies the seriousness of the skeptical position. The famous philosophical skeptics, after all, have gone to great lengths to publish their skeptical treatises and in doing so have made clear that they assume that there are other minds (and so implicitly other bodies). Skeptics who managed to survive made the same inductive inferences rejected as illegitimate in their skeptical "mode." They also argued with their fellow philosophers, past and present, and in doing so placed complete faith in the existence of a past revealed to them through memory. The very activity of philosophy, Butchvarov argues, is one that presupposes the existence of an external world, a past, and other minds, and the philosophers who realize this can take philosophy seriously only if they reject strong skepticism.[15] Reaching a skeptical conclusion is incompatible with taking oneself and one's work seriously. Philosophical skepticism is not then a serious philosophical position.

The charge invoked here is *not* the charge of formal epistemic self-refutation. The claim is not that the skeptic's conclusion entails that the skeptic has no reason to believe the conclusion. The alleged self-refutation is more subtle. It amounts to the claim that in embracing a skeptical conclusion as a serious philosophical position the skeptic is implicitly engaging in behavior that makes sense only against the backdrop of a set of beliefs that are incompatible with radical skepticism.

The first step in a skeptical response to this sort of criticism involves the obvious distinction between believing something and rationally believing something. Of course, there are no skeptics who withhold belief with respect to the questions of whether there is a past, a physical world (in some sense of "physical"), other minds, and regularities that can be safely projected into the future. But it is by no means obvious that to believe *P* one must believe that it is epistemically rational to believe *P*. I know all sorts of very religious people who seem to quite happily concede that their theism is epistemically irrational. Earlier, when discussing the question of whether the concept of justified belief is a normative concept, we very briefly raised the question of the extent to which belief is something we *control*. I sidestepped the issue of whether it was in principle possible to produce belief through an act of will, but it is difficult to deny that as a matter of empirical fact we simply find ourselves believing all sorts of things. It is not until our first philosophy class that many of us even raise the question of whether the things we take for granted are epistemically rational for us to accept. Indeed, it seems to me that philosophers too often forget that the questions raised by the skeptic are *philosophical* questions. They are questions raised by people who have a certain kind of philosophical curiosity that arises naturally from a very unnatural kind of activity. I argued earlier that there may well be different concepts of epistemic rationality and that some of these concepts might have particular relevance to *philosophy*. It may be that the philosopher is interested in and *wants* a kind of justification that ordinary people do not even think about in their day-to-day lives. The philosophical skeptic may best be construed as telling the philosopher that this kind of justification is unavailable. In every other walk of life people must get used to the idea that they cannot have everything they want, and the skeptic might maintain that it is a kind of perverted optimism to suppose that the *kind* of justification that would satisfy the *kind* of curiosity that afflicts the epistemologist is there to be found. This is a theme to which I return in the final chapter.

Notes

1. Klein (1981) calls this *iterative* skepticism.
2. Of course, when the necessary truths and falsehoods and the logical entailments get complex, the story may be different. In particular, if time is required to ''see'' the truth or entailment, then memory may be involved in the discovery of such truth or entailment. If memory is required to discover complex necessary truth or entailment, then any skepticism with respect to memory will infect all knowledge that explicitly or implicitly relies on it.
3. Although not neurophysiologists, the Martians in Bradbury's *The Martian Chronicles* caused the initial visitors from Earth no end of trouble through their telepathic ability to induce hallucinatory experiences. At the risk of exposing the shallowness of my aesthetic sensibilities, I would also mention the film *Total Recall*, whose plot centers on the question of whether the protagonist is really on a trip to Mars or is enjoying the experiences of an artificially induced fantasy.
4. Sartre (1966, part 4) would certainly reject the claim that our beliefs about others always involve inference if this is intended to be a phenomenological description of how our minds work. On the other hand, with philosophers in this tradition it is notoriously difficult to get an unambiguous statement of their views about the epistemological significance of their phenomenological observations.
5. In contrast, for example, to the more pivotal role they play in, say, Stroud 1984.
6. Indeed, some—Gilbert Harman, for example—have argued that inductive reasoning is itself a species of reasoning to the best explanation. See Harman 1965, 1970.

7. For a detailed discussion of this move by the skeptic, see Alston 1993, chapt. 4.

8. I argue in chapter 6 that it depends on the plausibility of externalist metaepistemologies.

9. Plantinga (1993b, chap. 12) appears to argue that it is difficult to see how natural selection would favor even true beliefs. The right false belief coupled with the right odd desires will do wonders for increasing the probability of my survival.

10. Hume 1888, p. 187.

11. Putnam 1981, chap. 1.

12. Fumerton and Foley 1985.

13. See Hume 1888, p. 210.

14. In his famous opening remarks in Meditation I.

15. See Butchvarov 1992.

Chapter Three

Internalist and Externalist Foundationalism

The traditional debate over skepticism has largely presupposed the framework of foundationalism. With the rise of the internalism/externalism debate in epistemology, however, it is apparent that there are radically different ways to understand foundational justification. In this chapter we begin by examining the traditional epistemic regress argument for foundationalism. Before presenting what I take to be the most important traditional conception of foundational justification, I examine the internalism/externalism controversy, or more appropriately, *controversies*, so that we may better understand the sense in which traditional foundationalism is (and is not) committed to internalism. This paves the way for a detailed examination in the next chapter of paradigm externalist versions of foundationalism. Having distinguished importantly different senses in which views about justification may be internalist or externalist, I examine in some detail what I take to be the most plausible traditional account of foundationalism, an account that is standardly regarded as internalist. As we shall see, however, one must be very careful to distinguish the senses in which this view is and is not committed to internalism. I conclude by distinguishing a *conceptual* from an *epistemic* regress argument for foundationalism.

The Principle of Inferential Justification and the Epistemic Regress Argument for Foundationalism

We saw in the last chapter that the skeptic relies heavily on the principle of inferential justification to support strong local skepticism with respect to various sorts of propositions. Ironically, perhaps, that same principle is assumed, together with a *rejection* of skepticism, to argue for classic foundationalism. The foundationalist holds that every justified belief owes its justification ultimately to some belief that is noninferentially justified. One wants to be able to present the argument for foundationalism without yet having a philosophically detailed analysis of what noninferential justification might consist in, and so it is helpful to begin with a neutral—and for that reason not very informative—characterization of the distinction between inferential and noninferential justification. Let us say that a belief that *P* is inferentially justified if its justification is constituted by the having of at least one belief other than *P*. A belief is noninferentially justified if its justification does not consist in the having of any other beliefs. This way of making the distinction leaves open the question of whether there are any noninferentially justified beliefs, and of course it leaves open the question of what might make a belief noninferentially justified.

Some paradigm internalists *and* almost all paradigm externalists embrace foundationalism. Both recognize a distinction, in principle, between inferentially and noninferentially justified beliefs, and both hold that all justified beliefs can trace their justificatory ancestry back to noninferentially justified beliefs. They hold radically different views as to what noninferential justification is. The classic epistemological alternative to foundationalism is the coherence theory of justification, a view that can be held in either an internalist or an externalist form, although, as we shall see, it is unlikely that the view will have much attraction in its externalist form. Part of what makes the contemporary metaepistemological scene so interesting is that such odd alliances are being formed. Classic foundationalists unite with coherence theorists in their opposition to externalism. Classic foundationalists and externalists unite against the coherence theorist in their support of foundationalism. Part of what we must do in the remainder of this chapter is to make clearer just exactly what makes someone an internalist or an externalist in epistemology.

As I indicated, the principle of inferential justification plays an integral role in the famous regress argument for foundationalism. If all justification were inferential, the argument goes, we would have no justification for believing anything whatsoever. If all justification were inferential, then to be justified in believing some proposition *P* I would need to infer it from some other proposition *E1*. According to the first clause of the principle of inferential justification, I would be justified in believing *P* on the basis of *E1* only if I were justified in believing *E1*. But if all justification were inferential I would be justified in believing

E1 only if I believed it on the basis of something else *E2*, which I justifiably believe on the basis of something else *E3*, which I justifiably believe on the basis of something else *E4*. . . , and so on ad infinitum. Finite minds cannot complete an infinitely long chain of reasoning, so if all justification were inferential we would have no justification for believing anything.

This argument relies on the first clause of the principle of inferential justification and is perhaps the most familiar version of the regress argument for foundationalism. It is important to realize, however, that we can invoke the second clause of the principle together with the first clause to generate not one but an infinite number of infinite regresses that face the antifoundationalist anxious to avoid global strong skepticism. To be justified in believing *P* on the basis of *E1*, we must be justified in believing *E1*. But we must also be justified in believing that *E1* makes probable *P*. And if all justification is inferential, then we must justifiably infer that *E1* makes probable *P* from some proposition *F1*, which we justifiably infer from some proposition *F2*, and so on. We must also justifiably believe that *F1* makes probable that *E1* makes probable *P*, so we would have to infer that from some proposition *G1*, which we justifiably infer from some proposition *G2*, and so on. And we would have to infer that *G1* makes probable that *F1* makes probable that *E1* makes probable *P*. . . . The infinite regresses are mushrooming out in an infinite number of different directions. If finite minds should worry about the possibility of completing one infinitely long chain of reasoning, they should be downright depressed about the possibility of completing an infinite number of infinitely long chains of reasoning. I call this the *epistemic* regress argument for foundationalism in order to distinguish it from what I take to be an even more fundamental *conceptual* regress argument for foundationalism, an argument I discuss later.

It should be noted that the epistemic regress argument for foundationalism requires the additional premise that strong global skepticism is false. In principle it is possible for a skeptic to argue that there is no such thing as a noninferentially justified belief and to use the regress argument as an argument for strong global skepticism. As we noted, strong global skepticism is the paradigm of a conclusion that will make any argument for it epistemically self-defeating, and the vast majority of skeptics are unwilling to embrace a conclusion this radical. Almost all skeptics are local skeptics who either explicitly or implicitly embrace the existence of noninferentially justified beliefs.

It is an understatement to suggest that the epistemic regress argument for foundationalism is not without its critics. As you might expect, the main criticism comes from the coherentists, who reject the crucial presupposition of the argument that justification is *linear* in structure. According to the coherentists, all justification always involves reference to other beliefs. The coherentist may also accept a version of the principle of inferential justification. The coherentist may well allow both that the other beliefs that justify me in believing *P* will themselves be justified and that for them to justify me in accepting *P* I must be aware of the way in which they bear upon the truth of *P*. The first clause of the principle of inferential justification does not generate a regress, however, because although my belief that *P* will depend on my having another justified belief, for example *E*, the justification for *E* will in turn rest in part on my having the belief that *P*. To be fair, one must stress the "in part." Coherentists are sometimes accused of endorsing the legitimacy of circular reasoning, but one does not really move in a circle at all. To caricature the view this way is to try to describe once again the coherentist's view of justification from within the confines of a linear conception of justification. One does not *first* justify one's belief that *P* by appeal to *E* and *then* justify one's belief that *E* by appeal to *F*, which one ultimately justifies by appealing once again to *P*. It is the entire structure of our belief system that justifies the individual beliefs that make it up. Each belief is justified by the relation it bears to the rest. Whether one can avoid the regress generated by the second clause of the principle is a subject to which we return in chapter 5 when we discuss the coherence theory of justification in detail.

The coherence theorist tries to avoid both the regress and foundations by rejecting the linear conception of justification and by rejecting the idea that the justification of justifiers must exclude reference to the belief being justified. A more heroic, albeit somewhat desperate, attempt to avoid foundations would be to acknowledge the existence of the regresses but deny that they are vicious. Philosophers, after all, tend to be somewhat paranoid about regresses. Not all regresses are problematic, and the fact that we are finite minds does not mean that we have a finite number of beliefs. Indeed, if one makes the standard distinction between occurrent and dispositional belief, it seems entirely plausible to suppose that we *do* have an infinite number of beliefs. Without analyzing the distinction yet, it seems a virtual truism that we think of ourselves (and others) as having beliefs in propositions which we are not currently considering. When I describe Goldman as believing that reliably produced beliefs are justified beliefs, I am not suggesting that right this very moment he is mulling over the hypothesis while nodding his head in agreement. In whatever sense it is that people can be said to believe propositions they are not currently considering, it seems entirely plausible to say of you that you believe that $2 > 1$, that $3 > 1$, that $4 > 1$, and so on ad infinitum. When you believe P, you also believe $P \vee Q$, $P \vee Q \vee R$, $P \vee Q \vee R \vee S$, and so on ad infinitum. Furthermore, it seems intuitively clear that all of these beliefs will be equally justified and thus that there is no obstacle to having an infinite number of justified beliefs. But if it is in principle possible to have an infinite number of justified beliefs, why should we assume that we would ever run out of beliefs to be the links in these infinitely long chains of reasoning?

It seems to me true that we not only can but do have an infinite number of justified beliefs. Nevertheless, one can plausibly argue that this will not avoid the need for foundations if we are going to reject strong global skepticism from within a framework presupposing that justification has a linear structure. The first problem is that the most obvious examples of someone's having infinitely many justified beliefs are examples where we can, in fact, trace the justification for these infinitely many beliefs back to a single proposition believed. I believe (perhaps justifiably) that it is a law of nature that all metal expands when heated. There is a sense in which I would also automatically be-lieve that if a were metal and heated it would expand, if b were metal and heated it would expand, and so on, but only because I realize that these propositions follow from the hypothesis that it is a law that all metal expands when heated. If we must employ this sort of model to understand the way in which we can come to have an infinite number of dispositional beliefs, then it is plausible to argue that when we have an infinite number of justified beliefs their justification is always ulti-mately owed to a finite set of beliefs. And when we get back to the finite set of beliefs, we will no longer have an infinite number of justi-fied beliefs upon which to draw in our search for justification. The infi-nite number of beliefs I have are always farther along the chain of linear justification and cannot be used as justifiers *prior to* the justification of the beliefs upon which their own justification is parasitic.

Although it is hardly an argument, another reason for doubting the possibility of avoiding foundations by embracing linear infinite re-gresses as harmless is the sheer inability to imagine how the regress would be continued. To be sure, foundationalists do not agree either on what noninferential justification consists in or on which beliefs are noninferentially justified, but it is terribly difficult to even imagine how one might continue to appeal to still more and more beliefs in justifying one's belief that one is in pain now. That is not to deny that philoso-phers have tried to find all sorts of things to which one not only can but

must appeal in order to have an epistemically rational belief that one is
in pain, but, these arguments typically confuse metabeliefs about the
correct linguistic descriptions of pain with beliefs about pain.

Yet another way in which one can avoid commitment to foundations
is to reject the principle of inferential justification upon which the regress argument relied. Rejection of the principle is not, however, a sufficient condition to avoid the regress argument. Externalists, we noted
in the last chapter, typically reject the second clause of the principle
but embrace the first, and for this reason are still likely to embrace
foundationalism.

If we decide that there are noninferentially justified beliefs and that
inferentially justified beliefs owe their justification ultimately to noninferentially justified beliefs, then we obviously need an account of both
noninferential and inferential justification. I suggested earlier that classic foundationalists and most paradigm externalists accept a version of
foundationalism, but before we distinguish internalist from externalist
versions of foundationalism we need a clearer understanding of how to
understand the internalism/externalism controversy.

Internalism/Externalism Controversies

I have made a number of references to the internalism/externalism
metaepistemological controversy without having yet defined it. Despite
the fact that much of contemporary epistemology takes place in the
shadow of the internalism/externalism debate, and despite the fact that
the controversy seems to strike deep at the heart of fundamental epistemological issues, I am not sure that it has been clearly defined. I fear
that philosophers are choosing sides without a thorough understanding
of what the respective views entail. In subsequent discussion I distinguish at least four different controversies that can be associated with
what is labelled the internalist/externalist debate in epistemology.

"Internal State" Internalism

Perhaps the most natural way of understanding the controversy, given
its name, is to define the internalist as one who is committed to the
view that epistemic properties are *internal* characteristics of a subject.
The internalist, on this view, maintains that S's knowing that P, or having a justified belief that P, consists in S's being in some internal state.
The externalist, by contrast, is one who is committed to the view that

knowledge and justified belief at least *involve* external states. Thus, on this way of defining the controversy, one can easily understand a common theme among internalists about *epistemic justification*, namely, that two people cannot be in identical internal states while one is justified in believing *P* and the other is not justified in believing *P*. The externalist, on the other hand, embraces the view that you and I can be in the same present internal states while one of us has a justified belief and the other does not. On many popular versions of externalism, for example, the epistemic status of one's belief depends crucially on its causal ancestry. Even though your present state is identical to mine, you might be justified in believing *P* because your present state came about in the "right" way, while I am unjustified in believing *P* because mine did not.

Anyone who wants to define the internalism/externalism controversy this way owes us an account of precisely what the contrast is supposed to be between internal and external states. Can my present internal states include relational properties? If they cannot, how many epistemologists are really prepared to identify the conditions that constitute having a justified belief with purely nonrelational properties? Paradigm internalists include sense-datum theorists who hold that it is direct acquaintance (a relation) with sense data that constitutes some kinds of noninferential justification. If internal states do include relational properties, it will be very difficult to draw the line between internal and external relational properties. Suppose, for example, that I am a classic foundationalist who embraces some version of direct realism. I hold that one can be directly acquainted with surfaces or other logical constituents of *external* objects and that such acquaintance is what justifies me in believing various propositions describing the external object. Such a view would be for many a paradigm of the kind of internalism under attack by contemporary externalism. But can we say that being acquainted with something external is itself an *internal* state of the subject who bears that relation of acquaintance to the *external* object? If one allows that internal states include relations whose relata include external objects, on the other hand, why should we deny that being presently caused by such and such a stimulus is an internal state? If we do not, of course, there is nothing to prevent certain paradigm forms of externalism (versions of a causal theory of justification) from moving into the internalist camp.

I should perhaps stress again that I do not think it makes any sense to suppose that there is some *one* way of defining the internalism/externalism controversy. We are dealing with a technical philosophical distinction, and our only concern should be with finding a way of characterizing the distinction that makes it illuminating and valuable by way of contrasting importantly different views. One certainly can define the controversy in terms of whether or not internal states of a person are sufficient to justify that person in believing various propositions. And one can formulate precise definitions of internal states. If one appreciates the difficulties one will encounter by allowing as internal states relational properties involving relata that are external to the subject, one could simply define internal states as nonrelational properties of the mind and those relational properties of the mind whose relata are themselves nonrelational properties of the mind. And if we define internalism this way, we can still find philosophers who will satisfy this definition of the internalist. If, for example, one embraces an adverbial theory of consciousness and holds that being appeared to in a certain way (perhaps coupled with thinking in a certain way) is sufficient for being justified in believing certain propositions about the physical world, one will be an internalist in the sense we just defined, at least an internalist with respect to this sort of justification. Or if one holds that it is acquaintance with nonrelational properties of the mind that yields justified belief, one will again satisfy the above characterization of an internalist with respect to this sort of justification. I personally do not think that the above definition gets at the heart of the controversy that divides contemporary epistemologists. I suspect that most people who think of themselves as externalists, for example, want to keep Moore, who contemplated the possibility that sense data have an existence external to and independent of the mind, in the internalist camp.

It is worth noting in passing that if one defines the controversy in terms of whether or not same internal states yield same epistemic status for beliefs (on virtually any plausible interpretation of ''internal''), there will be very few epistemologists who are internalists about *knowledge*. All those who hold a justified true belief account of knowledge, where the truth condition is nonredundant, are introducing into the analysis an external element. A belief's being true will constitute an external element unless one embraces a very radical coherentist account of truth that relativizes truth to the ''internal'' belief system of the individual believer.[1]

"Access" Internalisms

Perhaps as common, or even more common, an approach to understanding the internalism/externalism controversy involves focusing on the question of whether "access" to the conditions that constitute epistemically justified or rational belief is a necessary condition for the belief's being epistemically rational. Thus, many paradigm internalists seem to hold that a set of conditions X can constitute your justification for believing P only if you have access to the fact that X obtains *and* access to the fact that when X obtains the belief is likely to be true.[2] (This approach to defining internalism may not in the final analysis be so very different from the attempt to define the internalist as one who locates justification "in" the mind. When the Modern philosophers distinguished what was "in" the mind from what lay "outside" the mind, it is not implausible to suppose that at least sometimes the distinction was *epistemic* in nature. On one interpretation, for example, sense data were "in" the mind for a philosopher like Berkeley primarily in the sense that they were "before" the mind. That which is before the mind is that to which one has a certain direct or privileged access.)

I cannot stress too strongly here the importance of distinguishing a general access requirement for justified belief from the principle of inferential justification. The principle of inferential justification was just that—a principle concerning *inferential* justification. It maintained that when you use one *belief* to support another belief you must be justified in accepting the supporting belief and justified in believing that the truth of the supporting belief would make probable the truth of the supported belief. The principle of inferential justification does *not* assert that when one is inferentially justified in believing P one must have access to the fact that one is justified in believing P. Given what the principle of inferential justification explicitly states, it is entirely possible for me to satisfy the conditions it sets forth for being inferentially justified in believing P without satisfying the conditions required for being justified in believing that I am justified in believing that P.

The crudely worded access requirement, as I presented it above, is a *general* principle about justification. It applies both to inferential and to noninferential justification, and it is a very dangerous principle to accept because it clearly invites the specter of a vicious regress. Before elaborating, however, it will be useful to distinguish a *strong* from a *weak* access requirement. Let us say that strong access internalism with respect to epistemic justification maintains that in order for S to be justified in believing P, S must actually have access to the conditions that constitute that justification. The weak access internalist argues only that in order for S to be justified in believing P, S must be *able* to access the conditions that constitute his justification.[3] Alston (1989) embraces a weaker access requirement still, arguing that one should require only that a *ground* of justified belief be the *kind* of thing to which one has potential access. Alston's access requirement is doubly weak in that he explicitly rejects any requirement that one be able to access the *connection* between the ground of one's justification and the truth of the belief for which it is a ground. Since the adequacy of the ground is part of what makes it a justificatory ground, Alston in effect requires no access, real or potential, to the fact that one has justified belief.[4]

Now if one were cynical one might suspect that the common use of the term "access" by internalists who support conditions of access for justification is a way of trying to disguise the fact that one is really talking about knowledge or rational belief about rational belief. "Access" is itself an epistemic term. One has access to the fact that P only if one knows that P or perhaps has an epistemically rational belief that P. Thus, the strong access internalist with respect to epistemic justification is really advancing the view that in order to have an epistemically justified belief one must know or at least have an epistemically justified belief that one has an epistemically justified belief. For the purposes of the discussion let us suppose that it is the latter view, that "access" is to be spelled out in terms of "epistemically justified belief." There is no way of shutting down the requirement once it is introduced, and consequently S's having an epistemically justified belief that P will entail that S has an epistemically justified belief that he has an epistemically justified belief that P, and will entail that S has an epistemically justified belief that he has an epistemically justified belief that he has an epistemically justified belief, and so on ad infinitum. In one respect, the infinite regress encountered appears to be worse than the epistemic regress that foundationalism was designed to avoid (see the introduction to this chapter). To be justified in believing P, not only would one need an infinite number of beliefs, but one would need an infinite number of beliefs of ever-increasing complexity. I myself have a terrible time even keeping straight the proposition I am supposed to be believing when I move past the second or third metalevel. I shall return to a more careful analysis of whether and how regresses generated by access requirements are vicious in the section on the acquaintance theory later in this chapter.

Weak access internalism might seem to fare better when it comes to avoiding a vicious regress because it does not require that anyone actually have these increasingly complex metabeliefs. Weak access internalism, however, is also vague precisely because of the ambiguities of the modal operator. The *potentiality* presupposed in talk of potential access could be logical, lawful, or one of the many common but imprecise notions of possibility that float around in ordinary discourse. Let us say that *P* is logically possible if it involves no logical inconsistency. *P* is lawfully possible if the supposition that *P* occurs does not violate any laws of nature. As I indicated, there are, of course, other looser senses of possibility. To my great sorrow, I cannot dunk a basketball. It is not logically impossible for me to dunk a basketball, and my dunking a basketball would not violate any laws of nature. I cannot do it only in the sense that the present state of my body, the earth, the atmosphere, and the like, together with the laws of nature, preclude my dunking a basketball *in relatively normal conditions.* The normal conditions clause is needed to take care of freak updrafts, for example, that might enable me just this once to reach over the hoop.

Now the weak access requirement, it seems to me, is unlikely to get at the core of the internalism/externalism controversy. As we shall see in examining specific externalist views, it would be all too easy for an externalist to incorporate weak access requirements into an account of justified belief while retaining its essential externalist character. I argue that all of the paradigm externalists will (or at least should) think that it is logically possible to access the conditions that constitute justification, and most often they will think that it is lawfully possible to access the conditions for justification. They will not typically think that it is possible in the looser sense described above to access the conditions that constitute justification, but there is nothing in their externalism that would prevent them from adding such access conditions if they get tired of hearing internalists whining about the lack of access requirements.[5] I am not arguing that externalists would be sympathetic to the claim that *S*'s being justified in believing *P* entails *S*'s having the potential to access the conditions that constitute his being justified in believing *P*. I am only pointing out that an externalist who accepted such a view would stay an externalist *provided that the access referred to was still given a paradigmatic externalist understanding.*

One can try to use access requirements in order to distinguish in principle between internalist and externalist views by adding further restrictions to the way in which it must be possible to access the conditions that constitute justification. Thus Chisholm (1989) appears to argue that when one's belief has a certain epistemic status, it must be possible to discover that it has such a status by looking inward. If one holds that all and only internal states can be introspected, one might think of such a view as combining "internal state" internalism with "weak access" internalism. Of course, externalists have their externalist analyses of introspective knowledge, and it is still not clear that we are going to count as an internalist the philosopher who pays lip service to access requirements from within an externalist framework. Whether one adopts strong or weak access requirements for justified belief (even access requirements as weak as Alston's), one must be very careful to avoid vicious *conceptual* regress. If access is understood in terms of justified belief (or knowledge which in turn is analyzed as justified belief), one is clearly asking for trouble if one analyzes *all* justified belief in such a way that access requirements are introduced *into the analysis.* Such analyses will be transparently circular. The access requirements in one's analysis of justification will presuppose an understanding of the very epistemic concepts one is attempting to analyze. The solution for "access" internalists is to deny that their access requirements are *constitutive* of all justified belief even if actual or potential access to constituents of justification is in some strong sense *implied* by the nature of those constituents. I return to this point in the section on acquaintance.

Internalism and the Rejection of Naturalistic Epistemology

My own view is that the heart of the internalism/externalism controversy has very little to do with justification being an internal state or with the question of whether S's being justified in believing P implies that S is or has the potential to be justified in believing that S is justified in believing P. In fact, I think it has to do more with the older issue of whether or not to "naturalize" epistemology. I argue that the best way to think of the common thread that runs through paradigm externalist accounts of knowledge and justification is their reliance on the *reducibility* of epistemic concepts to nomological concepts. Thus the concept of reliability that is the building block of Goldman's account of justified belief seems clearly to be a concept that relies heavily on our understanding of nomic connections (whether we understand reliability in terms of causal propensities, probabilistic laws, or close causally possible worlds).[6] Armstrong (1973), of course, explicitly invokes causal concepts in explicating basic (foundational) beliefs that operate as that thermometer registering the temperature. Nozick's (1981) tracking relations, too, can only be explicated using causal concepts. To be sure, he employs the metaphor of close possible worlds in explaining the truth conditions of the counterfactuals that define tracking, but, to his credit, he realizes that the metaphor is just that—a metaphor. The contingent counterfactuals that define tracking assert nonlogical, nomic connections between the states of affairs referred to in the antecedents and consequents. The concept of information processing on which Dretske (1981) so heavily relies as the cornerstone of his approach to understanding justification seems just as clearly a concept that relies on an understanding of certain sorts of causal chains as the way to make clear the concept of a justified belief.

The paradigm internalist, by contrast, thinks that there are certain sui generis epistemic concepts that cannot be reduced to any more familiar, scientifically ''respectable'' concepts. These epistemic concepts may involve internal states but, as we shall see, this is a complicated issue. The clarity and distinctness of ideas that ground justification in self-evident truths for Descartes, the indefinable epistemic concept that Chisholm (1989) uses to define other epistemic concepts, the acquaintance that Russell (1926, 1959), Price (1950), and Hume (1888) all explicitly or implicitly appeal to in explicating noninferential justification, are clearly concepts that cannot be analyzed employing the concepts of causal or lawful (universal or probabilistic) contingent connections.

It should be obvious that if we define internalists and externalists in terms of their willingness to ''naturalize'' epistemic concepts by reducing them to scientifically ''respectable'' nomic concepts, we must recognize that there are important differences within the respective camps. The specific account of noninferential justification that I defend (and that I think is historically very important) relies heavily on an unanalyzable concept of acquaintance. Other ''nonnaturalists'' in epistemology invoke different primitives. In other words, given this way of understanding the internalism/externalism distinction, there will be different *versions* or *species* of both internalism and externalism (just as there are importantly different versions, for example, of both physicalism and dualism with respect to the analysis of mental states).

''Inferential'' Internalism

Earlier I suggested that there is an obvious disagreement between paradigm internalists and externalists over the plausibility of the principle of inferential justification. I have emphasized the importance of distinguishing acceptance of the principle of inferential justification from strong access internalism. But we might focus on the externalist's rejection of the second clause of the principle to characterize a distinction between what I call *inferential internalism* and *inferential externalism*. The inferential internalist believes that in order for me to be justified in believing *P* on the basis of *E*, I must be justified in believing that *E* makes *P* probable. The inferential externalist denies this. I noted earlier that the principle of inferential justification is not self-evidently a meta-epistemological view, but in the context of this metaepistemological debate let us say that the inferential internalist holds, whereas the inferential externalist denies, that S's justifiably believing that E makes P probable is literally a constituent of S's being inferentially justified in believing P on the basis of E. Since this analysis invokes the concept of justification, it can only be construed as partially analyzing the concept of *inferential* justification. The internalist who adopts this analysis of inferential justification will be tempted to borrow a leaf from the book of paradigm externalism and offer a recursive analysis of justification.

Summary

I have sketched some of the main ways in which one might try to understand the internalism/externalism controversy. Earlier I argued that in general it is unwise to suppose that there is only one way of correctly analyzing even those concepts that find expression in ordinary discourse. If it is a mistake to insist that there is only one analysis of familiar concepts, it is an even more obvious mistake to suppose that there is only one way of understanding technical philosophical terminology. There are really only two criteria one might use in choosing one way of understanding the controversy over another. First, we have *paradigm* representatives of the opposing camps, and I assume we want to understand the controversy so that these paradigm internalists and externalists get put in their respective camps. Goldman, Nozick, Dretske, and Armstrong are externalists.[7] Descartes, Hume, Chisholm, Price, and the Russell who emphasized the importance of direct acquaintance are internalists.[8] I want to understand the controversy so that these philosophers stay, respectively, externalists and internalists. Second, and more important, we want to understand the controversy in a way that makes it most interesting. We want to see what the *fundamental* disagreement is between the paradigm internalists and the paradigm externalists.

I introduced this preliminary discussion of the internalism/externalism debate because I did not want to begin an examination of internalist versus externalist analyses of noninferential justification in a vacuum. At the same time, I do not want to complete our discussion of how best to define the fundamental nature of the controversy until we have a chance to look at some of the prominent internalists and externalists, as I shall do now, beginning with internalist foundationalism. As we examine each view we shall try to keep clear the sense in which its proponent is or is not an internalist. Doing so should further clarify the implications of different forms of internalism and externalism.

Traditional Accounts of Noninferential Justification

Infallible Belief

If we look at some of the classic attempts to explicate the concept of noninferentially justified belief, we find that noninferential justification is often implicitly or explicitly identified with infallible belief. It seems fairly clear, for example, that Descartes wanted his foundations for knowledge to consist in beliefs that could not possibly be mistaken. When Price introduced the notion of sense data, knowledge of which would be included in the foundations of empirical knowledge, he contrasted sense data and their nonrelational properties with other sorts of things about which one could be mistaken, implying that the way to find the correct foundational knowledge is to scrape away from one's beliefs all that could be false. There are, however, a number of different ways of formulating the notion of an infallible belief. Lehrer (1974) initially suggests that we understand infallibility in terms of whether or not the occurrence of the belief entails the truth of what is believed:

Ia. S's belief that P at t is infallible if S's believing P at t entails that P is true.

Let us construe the entailment broadly so that P may be said to entail Q if P formally, analytically, or synthetically entails Q. (P formally entails Q if it is a tautology that if P, then Q; P analytically entails Q when the proposition If P, then Q can be turned into a tautology through the substitution of synonymous expressions; P synthetically entails Q when the proposition If P, then Q is true in all possible worlds but is neither a tautology nor an analytic truth.)

Lehrer is a critic of foundationalism, and one should beware of helpful suggestions from coherentists on how to search for appropriate foundations for empirical knowledge. Although Ia gives a perfectly intelligible definition of infallibility, it is far from clear that it has much relevance to an attempt to understand noninferential *justification*. The problems with associating infallibility and noninferential justification are familiar and I will only sketch some of them. First, there are the paradoxes of entailment recognized by Lehrer himself. Every necessary truth is entailed by every proposition, and thus if I happen to believe a necessary truth, *P,* the fact that I believe *P* will entail that *P* is true. Thus, by Ia my belief that *P* will be infallible whenever *P* is a necessary truth no matter how complex *P* might be. *P* could be a necessarily true mathematical proposition whose truth I could never ''see'' directly and whose proof I could never understand, but if I believed *P* because I always believe the first mathematical proposition I consider on Tuesdays, my belief that *P* would be infallible. Surely this concept of infallibility has precious little to do with whether or not my belief is justified.

To deal with this problem one can tinker (as Lehrer does) with the definition of infallible belief, but once one understands that mere entailment between the having of a belief and the truth of what is believed does not necessarily provide justification, one should be wary about assuming that it ever does, even in the few uncontroversial instances in which believing a contingent proposition entails that the proposition is true. The following beliefs are all infallible in the sense defined by Ia: my belief that I exist, my belief that I am in a conscious state, and my belief that I believe something. The first of these entailments appears to be the one that Descartes thought so significant in arriving at secure foundations. The entailments do hold, but is the mere existence of the entailment enough to provide anyone with justification? If it is not in the case of necessary truths entailed by beliefs, why should it be sufficient in the case of contingent truths entailed by beliefs?

Another standard objection to understanding foundational beliefs in terms of this concept of infallible belief is the alleged difficulty of finding *enough* empirical propositions that can be infallibly believed. Once one gets beyond the ''trick'' examples of belief in propositions whose content explicitly or implicitly refers back to the belief, the argument goes, there are no infallible empirical beliefs, and a foundationalism attempting to build on infallible beliefs understood in the above way is not going to support the edifice of justified beliefs we are trying to erect. Some philosophers attempt to establish this conclusion by considering specific candidates for infallible beliefs and describing elaborate hypothetical situations designed to convince us that we could reasonably decide that the beliefs in question were false. Thus Armstrong (1963) imagines us once again in a future characterized by a utopian neurophysiology. We are wired to complex machines that inform us that, despite the fact that we believe we are in pain, there is simply no indication of the sort of neural activity associated with pain. Might it not be reasonable in such a situation to conclude that we simply have a false belief that we are in pain? Lehrer (1974) argues that one can genuinely (as opposed to merely verbally) confuse pains with itches and for that reason arrive at a false belief that one is in pain.

We will return to these arguments shortly, but there is a very general argument designed to establish that the foundationalist's favorite candidates for noninferentially justified empirical beliefs are not infallible. It is a Humean sort of argument that proceeds from the simple observation that in the vast majority of cases, the belief that *P* is one state of affairs and *P*'s being the case is a different state of affairs. If these really are two distinct facts, then why couldn't one have the one without the other?[9] Although it does not add much to the logical force of the argument, one can again employ our hunches about how the brain might work to rhetorically bolster the argument. Consider again a standard candidate for an infallible empirical belief, my belief that I am in pain now, for example. It is surely *possible* that the region of the brain causally responsible for producing the belief that I am in pain is entirely different from the region of the brain causally responsible for producing the pain. There may be a causal connection between the occurrence of the "pain" brain event and the occurrence of the "belief" brain event, or vice versa, but even if the causal connection holds it will be a contingent fact that it does. It hardly seems that the neurophysiologist could discover these (or any other) causal connections purely a priori. But if the brain state responsible for my belief that I am in pain is wholly different from the brain state responsible for the pain, and if the connections between them are merely nomological, then it is in principle possible to produce the one without the other. The belief will not entail the truth of what is believed.

Infallible *Justification*

The foregoing argument has a great deal of plausibility, I think, and in any event it has always seemed strange to me to search for foundations in mere *belief.* What justifies me in believing that I am in pain? The mere fact that I believe that I am in pain? What *is* it about this belief that makes it so different from other beliefs? Why does my belief that I am in pain constitute a kind of justification but my belief that there are ghosts does not constitute a kind of justification? The appeal to belief as a justifier borders on a non sequitur if one is genuinely attempting to find a useful characterization of a special kind of epistemic relation one can bear to truth that obviates the need for inference.

As BonJour pointed out, this same lack of a genuine response seems to characterize those foundationalists who seek to identify the source of noninferential justification with the fact that makes the noninferentially justified belief true.[10] When asked what justifies one in believing that one is in pain, this foundationalist identifies the pain itself. What justifies me in believing that I am having a certain visual sensation is the visual sensation itself. But what is it about the pain or the visual sensation that makes it a justifier? When you believe that I am in pain, my pain doesn't justify you in believing that I am in pain (according to this foundationalist account), so there must be something different about my *relationship* to my pain that enters into the account of what constitutes the justification. It is the fact that I have a kind of access to my pain that you don't have that makes my belief noninferentially justified while you must rely on inference. One still needs an account of what this relation is, but before we consider such an account it is worth noting that we could have defined the concept of infallible belief in a way that makes it potentially more useful in developing a foundationalist theory of justification. The relevant question is not whether my belief entails the truth of what is believed. It is, rather, whether my *justification* entails the truth of what is believed:

Ib. *S*'s belief that *P* at *t* is infallible if *S*'s justification for believing *P* at *t* relevantly entails the truth of *P*.

It is necessary to qualify the entailment as relevant to circumvent the problems already discussed in connection with Ia. Whenever I have any justification at all for believing a proposition that turns out to be necessarily true, that justification will entail the necessary truth. But we do not want just any sort of justification to yield infallibly justified belief even if the object of that belief is a necessary truth. What is the difference between relevant and irrelevant entailment? This question is notoriously difficult to answer, but intuitively it should have something to do with the fact that would make true the proposition entailed and the fact that would make true the proposition that entails it. More specifically, we could say that *P* relevantly entails *Q* only if the fact that would make *P* true is at least a constituent of the fact that would make *Q* true. This suggestion can be considered at best only a preliminary suggestion, since we will obviously need a more detailed account of facts and their constituents. That I have grey hair entails that someone has grey hair, but is my having grey hair a constituent of the fact that is someone's having grey hair? There is certainly a sense in which it is something one can point to in answer to the question "What makes it true that someone has grey hair?" One cannot appropriately point to my having grey hair as something that makes it true that two plus two equals four. An enormous amount of work would have to be done in order to develop a coherent, plausible relevance logic, but these few comments might allow me to get away with using the concept of relevant entailment in this context.

Acquaintance and Noninferential Justification

I have suggested that neither a belief nor the truth of what is believed is by itself a plausible justification at all, let alone the kind of justification that might entail the truth of what is believed. Rather, we must stand in some sort of special *relation* to the truth of what is believed, or more precisely, we must stand in some sort of special relation to the fact that makes true what we believe. I have argued elsewhere that the most fundamental concept required to make sense of traditional foundationalism is the concept of acquaintance. In order to explain my acquaintance theory of noninferential justification, however, I must briefly digress and sketch a highly controversial theory of truth and intentionality.

I take the primary bearers of truth value to be thoughts (which I also refer to as propositions). The secondary bearers of truth value are the linguistic items that express them. Thoughts I take to be nonrelational properties of a mind or self, properties whose presence is logically distinct from, though no doubt causally dependent on, and paralleled by, brain states. Thoughts can be true or false. True thoughts correspond to or ''picture'' facts. False thoughts fail to correspond. A fact is a nonlinguistic complex that consists in an entity or entities exemplifying properties. The world contained facts long before it contained minds and thoughts. In one perfectly clear sense the world contained no truths before there were conscious beings, for without conscious beings there would be no bearers of truth value. There were facts that would have made true the relevant thoughts had they existed, and by employing counterfactuals we can make good sense of such commonplace assertions as that it was true hundreds of millions of years ago that there were no conscious beings.

Although I once thought the difference between believing, fearing, hoping, and other intentional states should be understood in terms of a relation that the mind bears to its thought, I now believe that every intentional state *is* a thought. Believing that there are ghosts, fearing that there are ghosts, and hoping that there are ghosts are all *species* of the same thought that there are ghosts. Believing and hoping that there are ghosts stand to each other as a blue and a yellow Ford Mustang stand to each other. We can represent true and false belief respectively as follows:

S believes truly that *P* = Df   '*P*'*s and '*P*' *C P*

S believes falsely that *P* = Df   '*P*'*s and it is not the case that
                                                                there exists some fact *x* such that
                                                                '*P*' *C x*

where *s* stands for *S*, '*P*' stands for the thought that *P*, * indicates that the thought is a belief, *C* stands for correspondence, *P* refers to the fact that *P*, and *x* is a variable.

This correspondence theory of truth avoids the need for such ontological nightmares as nonexistent states of affairs to serve as the "objects" of false beliefs, and it preserves a much more natural way of understanding the *referents* of sentences, analogous to the referents of names and definite descriptions. Unlike Frege, we have no need for such mysteries as The True and The False to serve as the referents of true and false sentences, respectively. Rather, we adopt the more straightforward view that just as the successful use of a name refers to an individual, so the successful—that is true—attempt to refer to the world with a descriptive sentence succeeds in picking out a fact. Some names, like "Pegasus," do not succeed in referring to any individual, and some sentences, like "Dogs have eight legs," do not refer to any fact. New theories of reference aside, having a referent is not necessary for having meaning, and the thoughts that false sentences express give those sentences meaning despite the fact that they fail to refer.

Acquaintance is *not* another intentional state to be construed as a nonrelational property of the mind. Acquaintance is a *sui generis relation* that holds between a self and a thing, property, or fact. To be acquainted with a fact is not *by itself* to have any kind of propositional knowledge or justified belief, and for that reason I would prefer not to use the old terminology of knowledge by acquaintance. One can be acquainted with a property or fact without even possessing the conceptual resources to *represent* that fact in thought, and certainly without possessing the ability to linguistically express that fact. But if this is true, what has acquaintance got to do with epistemology?

Sellars once argued that the idea of the given in traditional epistemology contains irreconcilable tensions. On the one hand, to ensure that something's being given does not involve any other beliefs, proponents of the view want it to be untainted by the application of *concepts*. The kinds of data that are given to us are also presumably given in sense experience to all sorts of other creatures. On the other hand, the whole doctrine of the given is designed to end the regress of justification, to give us secure foundations for the rest of what we justifiably infer from the given. But to make sense of making inferences from the given, the given would have to be propositional. Minimally, the given must have a truth value. But the kind of thing that has a truth value involves the application of concepts or thought, a capacity not possessed by at least lower-order animals.[11]

The solution to the dilemma presented by Sellars and others is to reemphasize that acquaintance is not *by itself* an epistemic relation. Acquaintance is a relation that other animals probably bear to properties and even facts, but it also probably does not give these animals any kind of justification for believing anything, precisely because these other animals probably do not have beliefs to begin with. Without *thought* there is no truth, and without a bearer of truth value there is nothing to be justified or unjustified. But how does acquaintance give us noninferential justification? My suggestion is that one has a noninferentially justified belief that *P* when one has the thought that *P* and one is acquainted with the fact that *P*, the thought that *P*, *and* the relation of correspondence holding between the thought that *P* and the fact that *P*. No single act of acquaintance yields knowledge or justified belief, but when one has the relevant thought, the three acts together constitute noninferential justification. When everything that is *constitutive* of a thought's being true is immediately before consciousness, there is nothing more that one could want or need to justify a belief.

The reader might well complain that if mere acquaintance with a fact does not constitute an epistemic property, surely one cannot conjure up an epistemic property by multiplying acts of acquaintance. But if this is intended to be a formal objection to the view I presented, it involves committing the fallacy of division. Because none of the components of a complex state of affairs constitutes the exemplification of an epistemic property, it does not follow that the complex does not constitute the exemplification of such a property. Classical acquaintance theorists like Russell appropriately emphasized the role of acquaintance with particulars, properties, and even facts in grounding justification. But a fact is not a truth, and what one needs to end a regress of justification is a direct confrontation with *truth*. To secure that confrontation, one needs to be directly aware of not just a truth-maker (a fact to which a truth corresponds) but also a truth-bearer (a thought) and the correspondence that holds between them.

Because the relations of acquaintance and correspondence that the above account appeals to are *sui generis*, there is precious little one can say by way of trying to explain the concept to one who claims not to understand it. Because acquaintance is not like any other relation, there is no useful genus under which to subsume it. One can give examples of facts with which one is acquainted and in this way present a kind of "ostensive" definition of acquaintance, but philosophers who think the concept is gibberish are unlikely to find themselves acquainted with their being acquainted with various facts. When one is acquainted with a fact, the fact is *there* before consciousness. Nothing stands "between" the self and the fact. But these are metaphors and in the end are as likely to be misleading as helpful. Correspondence, too, is sometimes thought of as a picturing relation, but the picturing metaphor is largely responsible for the caricature of the view one so often encounters in the cruder theories of "ideas" as pale copies of reality. It is tempting to at least mention the metaphor of a Kodak print and the scene it depicts as a way of explaining the relation that a true thought bears to the fact with which it corresponds, but most thoughts are not "pictures" and the relation of correspondence has nothing to do with any kind of similarity that holds between the thought and the fact it represents. Correspondence is not like anything else; it cannot be informatively subsumed under a genus, and it cannot be analyzed into any less problematic concepts.

Is acquaintance a source of infallible justification? The answer is in one sense straightforward. If my being acquainted with the fact that *P* is part of what justifies me in believing *P* and if acquaintance is a genuine relation that requires the existence of its relata, then when I am acquainted with the fact that *P*, *P* is true. The fact I am acquainted with is the very fact that makes *P* true. The very source of justification includes that which makes true the belief. In a way it is this idea that makes an acquaintance foundation theory so attractive. I have no need to turn to other beliefs to justify my belief that I am in pain because the very fact that makes the belief true is unproblematically before consciousness, as is the correspondence that holds between my thought and the fact. Again, everything one could possibly want or need by way of justification is there in consciousness.

Notice that the infallibility of the justification provided by acquaintance is due to the presence of the fact itself as a constituent of the justifier. It is interesting to note that in this respect there are remarkable similarities between this classic version of foundationalism and at least some paradigmatic externalist views. On certain causal theories of direct knowledge, for example, my belief that *P* is justified by its being caused in the appropriate way by the fact that *P*, the very fact that makes my belief true. If a causal relationship between the fact that *P* and my belief that *P* were a kind of justification, then that justification too would be infallible. Its existence would, trivially, entail the truth of what I believe. From the fact that a certain justification is infallible, it does *not* follow that one could not mistakenly believe that one has an infallibly justified belief. Certainly the causal theory I have just sketched would have no difficulty imagining a person who mistakenly concluded that his belief that *P* was caused by the fact that *P*, and if the causal theory were correct, that person could mistakenly infer that the justification in support of his belief entailed the truth of what he believed. Similarly, I think that it is in principle possible for a person to mistakenly conclude that he is acquainted with something actually known only through inference. One might trust a philosopher with a mistaken epistemology, for example, and falsely, perhaps even justifiably, believe that one is acquainted with a fact when one is not. Although this complicates matters considerably, I also argue that it may be possible on an acquaintance theory to have noninferential justification that does not entail the truth of what is believed. Specifically, I have argued that one might be acquainted with a fact very similar to the fact that makes *P* true, and such acquaintance might give one a justified but false belief that *P*. It should be clear that this admission is perfectly compatible with the rather trivial claim that when one's justification for believing *P* consists in part in being acquainted with the fact that *P*, that justification is infallible in that it entails the truth of *P*.

If I am asked what reason I have for thinking that there is such a relation as acquaintance, I will, of course, give the unhelpful answer that I am acquainted with such a relation. The answer is question-begging if it is designed to convince someone that there is such a relation, but if the view is true it would be unreasonable to expect its proponent to give any other answer. I can also raise dialectical considerations and object to alternatives. One of the dialectical advantages of the above view is that it can easily respond to some of the classic arguments against the existence of noninferentially justified belief.

One of the most discussed arguments against foundationalism again focuses on concepts. There is no truth value without concept application, the argument goes. But to apply a concept is to make a judgment about class membership, and to make a judgment about class membership always involves relating the thing about which the judgment is made to other paradigm members of the class. These judgments of relevant similarity will minimally involve beliefs about the past, and thus be inferential in character. Our reply to the argument is straightforward. To make a judgment, say that this is red, involves having the thought that this is red, but the thought does not involve relating this to some other thing. Indeed, it is in principle possible to produce a thought of red in the mind of someone who has never experienced a red thing. Since language is only a secondary and conventional means of representation, it goes without saying that the inferential character of our judgments concerning the linguistically correct way to express a thought are neither here nor there when it comes to the question of whether the thought expressed can be noninferentially justified.

The intelligibility of the above account does rest on the intelligibility of a world that has structure independent of any structure imposed by the mind. Without nonlinguistic facts that are independent of the thoughts that represent them, one could not make sense of a relation of acquaintance between a self and a fact, a relation that grounds direct knowledge. Indeed, I suspect that it is concern with this idea that lies at the heart of much dissatisfaction with traditional foundationalism. Since Kant there has always been a strong undercurrent of antirealism running through philosophy. The metaphor again is that of the mind imposing a structure on reality. And there is an intuitively plausible sense in which one can genuinely wonder whether it makes sense to ask about the number of colors that are exemplified in the world independently of some framework provided by color concepts. But despite the periodic popularity of extreme nominalism and rampant antirealism, it is surely absurd to suppose that it is even in principle possible for a mind to force a structure on a *literally* unstructured world. There are indefinitely many ways to sort the books in a library and some are just as useful as others, but there would be no way to begin sorting books were books undifferentiated. The world comes to us with its differences. Indeed, it comes to us with far too many differences for us to be bothered noticing all of them. And it is in this sense that the mind *does* impose order on chaos. Thought is abstract in the sense that many different actual properties can all correspond to a single thought of red. And it is up to us how finely we want to draw our color concepts. Although I understand that the empirical evidence is at best questionable, it is common for philosophers to call our attention to the alleged fact that some cultures have far more finely grained color concepts than our culture. If one distinguishes color concepts from linguistic terms to express those concepts, the empirical claim is difficult to assess, but one must surely admit that the alleged phenomenon is in principle possible. Given the above framework for understanding thought and truth, there would be a sense in which the one culture would entertain truths about colors that the other culture would be causally unable to accept. But the fact that there is good sense to be made of the relativity of conceptual frameworks should not mislead one into thinking that the properties exemplified in the world depend for their existence on concepts.

The Acquaintance Theory of Noninferential Justification and the Internalism/Externalism Debate

I have presented at length a view that I take to be the most plausible version of classical foundationalism. In what sense, if any, is it internalist? Well, on the specific version of the view I defended, thought is an internal property of the mind, if by "internal" one means "nonrelational." The crucial concepts of acquaintance and correspondence, however, are relational. It is true that given my own views in normative epistemology, it turns out that it is always a mental state or feature of a mental state with which we are acquainted, and so the complex act of being acquainted with X will involve constituents all of which are "internal" to the subject. But it should be emphasized that the *meta-epistemological* acquaintance theory of noninferential justification does not by itself entail any position with respect to what might be the objects of acquaintance. In previous discussion of the attempt to define internalism in terms of internal states being sufficient for justification, I noted that one might be a sense-datum theorist who thinks one can be directly acquainted with the fact that the surface of a physical object exemplifies a certain property. One might think that there are mind-independent universals and claim to be acquainted with them. One might think that there are mind-independent, nonoccurrent states of affairs and claim to be acquainted with logical relations that hold between them. It is at best unclear as to whether or not any of the above acts of acquaintance should be called internal states, and thus equally unclear as to whether a foundationalism defined using the concept of acquaintance is always going to be a species of "internal state" internalism.

As we saw earlier, being internal might also be understood in terms of access. There appears to be a historical use of "in" the mind which makes "in" an epistemic concept. When philosophers used to argue that sense data are "in" the mind, they may have sometimes meant only that we have a kind of privileged access to sense data. Does an acquaintance theory hold that when one is noninferentially justified in believing P one has access to—that is, knowledge or justified belief about—the fact that one has such justification? On the face of it, the answer seems to be no. In the paradigm case, I am noninferentially justified in believing P when I have the thought that P and am simultaneously acquainted with the thought that P, the fact that P, and the relation of correspondence holding between them. To have noninferential justification for believing that I am noninferentially justified in believing P, I must have that rather complex thought and simultaneously be acquainted with its correspondence to an equally complex fact. And for me to be noninferentially justified in believing that I am noninferentially justified in believing that I am noninferentially justified in believing that P, I must be acquainted with facts so complex as to boggle my poor consciousness. Indeed, I am not sure I can keep things straight past the fourth or fifth level. The position that in order to have a noninferentially justified belief on an acquaintance theory one *must* be noninferentially justified in believing that one has such justification invites a vicious regress of infinitely many, increasingly complex conscious states. I would strongly suggest that classic foundationalists decline the invitation.

In his influential attack on foundationalism, BonJour (1985) tried to entice the foundationalist into accepting what amounts to strong access internalism. His argument then involved claiming that any attempt to stop the regress of justification with a noninferentially justified belief would inevitably fail, because the justification would necessarily involve other beliefs. The argument was presented in a schematic way so that it applies to any version of foundationalism including the acquaintance theory discussed above. Let X be the properties of a belief that the foundationalist says constitute my noninferential justification for believing P. BonJour argues (pp. 31–32) that internalism entails that my belief's being X could give me justification for believing P only if I were (1) aware that my belief has these properties X and (2) aware that when a belief has properties X it is likely to be true. At least one of these propositions could be known only as a result of inference. (I argue later that BonJour is committed to the view that knowledge of probability relations is noninferential and so is really committed to the conclusion that it is the awareness of characteristics X that will require inference.)

If to be noninferentially justified in believing P requires that I have the justified *belief* that I am noninferentially justified in believing P, then it initially seems correct that there could not really be any noninferential justification. My belief that P, you recall, is noninferentially justified only if its justification does not consist even partially in the having of other justified beliefs. But my belief that I am justified in believing *P* is a different belief than the belief that *P*, and if the justified metabelief is required for the justification of the first-level belief, how can the first-level belief be noninferentially justified? Ultimately, I believe that the foundationalist must conclude that one cannot hold that noninferential justification entails being justified in believing that one has noninferential justification. Such a view does lead to a vicious regress. In discussing access requirements, however, I briefly distinguished two ways in which one might introduce such requirements, only one of which leads to *conceptual* regress. There is an important distinction between a belief's justification *entailing* the having of other justified beliefs and a belief's justification *consisting* in the having of other justified beliefs. In evaluating the nature of the regress generated by access requirements, we must keep this distinction firmly in mind.

The iteration requirement for noninferential justification is literally *unintelligible* if the access requirement is thought of as part of the *analysis* of justification. One is challenged to come up with a set of conditions *X* that *constitute* noninferential justification for believing *P*. One is then invited to accept the claim that those conditions constitute justification only if one adds awareness of them. But that is tantamount to admitting that *X* was not a satisfactory analysis to begin with. *The idea that X constitutes one's justification for believing P only if one's awareness of X is added to X is equivalent to holding that X constitutes one's justification for believing P only if it does not really constitute one's justification for believing P.* If I must embrace that conditional to be an internalist, I wash my hands of the view.

One can avoid this sort of *incoherent* access internalism by holding that *X* by itself constitutes justification and is therefore sufficient for justification, but that the very nature of *X* metaphysically entails that one has an awareness of *X*. On such a view the *analysis* of justification would not involve reference to second-level awareness of the conditions that constitute justification. Consider an analogy. *P*'s being true entails that it is true that *P* is true. In accepting this proposition I am *not* committed to including its being true that *P* is true in the analysis of what makes *P* true. In the same way, a strong access internalist could argue that it is necessarily true that the conditions that constitute justification always obtain together with conditions that constitute being justified in believing that they obtain even though the conditions that constitute the second-level justification do not belong in an analysis of the conditions that constitute first-level justification. Such a view avoids a conceptual regress (just as holding that *P*'s being true entails that it is true that *P* is true involves no conceptual regress). Unlike the regress of truth, however, the strong access requirement seems to give rise to a *vicious* regress. First-level justification entails the existence of second-level justification, which entails the existence of third-level justification, and so on ad infinitum. The metabeliefs required get more and more complex, and even if a finite mind can have an infinite number of justified beliefs, it is hard to see how it could have an infinite number of ever increasingly complex beliefs (supported by increasingly complex justification).

As I pointed out earlier, it is much easier to be a weak access internalist. One must still be careful to avoid the conceptual incoherence described above. One should not construe the possibility of access as a further condition in the *analysis* of justification. Nevertheless, one could argue that it is true, perhaps even necessary in some metaphysical sense, that when I am justified in believing *P* there is the possibility of access to that justification. I suggested earlier that if the possibility of access is understood broadly enough, the view is perfectly compatible with most versions of externalism. Whenever one has noninferential justification for believing *P*, I certainly think that it is logically possible that one has justification for believing that one is noninferentially justified in believing *P*. Furthermore, whenever one is noninferentially justified in believing that *P*, I think that it is in principle possible to have a noninferentially justified belief that one has a noninferentially justified belief. Of course, to have a justified belief at the second level would require possessing the requisite concepts, that is, it would require the capacity to think of such things as acquaintance and correspondence; and it is not certain that nonphilosophers possess such concepts—it's not even certain that most philosophers today possess such concepts. A *noninferentially* justified belief that one has a noninferentially justified belief that *P* would also require acquaintance with the complex set of acquaintances that constitute noninferential justification for believing *P*. And while I think that acquaintance is the sort of thing one could be acquainted with, I am not suggesting that whenever one is acquainted with something one is acquainted with the fact that one is acquainted with it. As we move up levels, it is not even clear to me that it is *causally* possible to become acquainted with the ever increasingly complex facts involving acquaintance that would be required to yield noninferentially justified beliefs.

If this version of what I am calling traditional foundationalism does not *entail* ''internal state'' internalism and denies strong access internalism and even some versions of weak access internalism, why call it internalism at all? The answer I suggested earlier points to its reliance on the sui generis concept of acquaintance that is fundamental to epistemology and that cannot be reduced to nonepistemic concepts, particularly the nomological concepts upon which all externalists build their analyses. I argued earlier that the heart of the internalist/externalist debate may have little to do with access requirements for justification (though it does have something to do with access to inferential connections). That never was the fundamental point of disagreement between internalists and externalists, and the proof of this is that externalists could in principle incorporate access requirements into their still *externalist* metaepistemological theories. Consider an odd sort of reliabilist, for example, who argues that one is noninferentially justified in believing that *P* when the belief that *P* is produced by a reliable process that takes as its "input" states that are not beliefs (I will discuss such a view in more detail shortly). And now suppose that our reliabilist decides that this is not enough, that one also needs "access" to the fact that the belief is produced this way. When asked what such access would consist in, the reliabilist responds that it would involve having a reliably produced belief that one has a reliably produced belief. And access at the third level would require having a reliably produced belief that one has a reliably produced belief that one has a reliably produced belief. Now no actual externalist is going to hold that justification at the first level requires that there be a justified metabelief about the justification of that first-level belief, and an attempt to incorporate such a view into one's externalism might well involve all kinds of insuperable problems including the potentially vicious regresses described earlier. But that is irrelevant to the question I am raising. However misguided an externalism with built-in access requirements might be, the theory will be no less externalist because of those access requirements if the access is still understood in causal terms.

The fundamental difference between externalism and one historically prominent and important form of internalism is that the internalist wants to ground all justification on a "direct confrontation" with reality. In the case of a noninferentially justified belief, the internalist wants the fact that makes true the belief "there before consciousness." The externalist can pay lip service to these desires by giving an externalist analysis of being confronted with a fact or having a fact before consciousness, but the internalist is convinced that no attempt to explain that immediacy in terms of *nomological* relations will succeed.

The above diagnosis of the essential difference between internalism and externalism is complicated by the fact that some externalists will deny that they are *reducing* epistemic concepts to nonepistemic nomological concepts. Goldman, for one, is clear about the fact that he is not trying to *define* epistemic terms "naturalistically."[12] Such definitions, he feels, would leave out the alleged normativity of the terms of epistemic evaluation. He is, nevertheless, attempting to provide necessary and sufficient conditions for a belief's being justified—he is trying to discover the conditions on which epistemic justification *supervenes*. There are as many different species of supervenience, however, as there are species of necessity and sufficiency, and it is clear that the kind of supervenience that Goldman thinks one discovers in a correct philosophical analysis is stronger than any sort of lawful connection. In terminology that is not popular with all philosophers, one might describe his conception of a philosophical account of justification to be the search for conditions that are synthetically necessary and sufficient for *S*'s being justified in believing *P*. And I would qualify the earlier discussion of the fundamental difference between traditional foundationalism and external foundationalism to emphasize that the traditional "internal" foundationalist denies that the fundamental confrontation with reality that yields noninferential justification can be reduced to, or even be viewed as strongly supervenient upon, nomological relations. (*X* strongly supervenes on *Y* only if in all possible worlds in which *Y* obtains *X* obtains.)

At the risk of being repetitive, I want to make clear one more time that I am happy to recognize that there is more than one way to define the internalism/externalism controversy. Indeed, I insist on recognizing importantly different interpretations of the controversy. I have tried to explain why I think it is most illuminating to characterize the *heart* of the controversy as a disagreement over the plausibility of naturalism in epistemology. I also emphasize again that the acquaintance account of noninferential justification I offered is only one version or species of an internalism that rejects naturalistic epistemology. Obviously, one might agree with me that the fundamental mistake of externalism is its attempt to reduce the epistemic to the nomological but deny that an understanding of epistemic concepts should be based (in part) on the sui generis concept of acquaintance. Chisholm, for example, uses a different primitive in advancing his version of internalism.[13]

I have tried to develop a plausible traditional foundationalist account of noninferential justification, and even before contrasting it with an externalist foundationalist account of noninferential justification, I have tried to indicate what I take to be the most illuminating sense in which it is an internalist view. I have argued that traditional foundationalists should be very careful about accepting strong and some weak access requirements. Earlier, however, I emphasized the need to make clear the distinction between accepting access requirements and accepting both clauses of the principle of inferential justification. The principle of inferential justification entails neither strong nor weak access internalism. It is a principle whose second clause will be rejected by almost all externalists, and by focusing on this fact we will be able to draw another important distinction between *inferential* internalism and *inferential* externalism. That metaepistemological difference will be revealed in their respective analyses of inferential justification.

Inferential Internalism, the Analysis of Inferential Justification, and a Conceptual Regress Argument for Foundationalism

The Principle of Inferential Justification

When I first introduced the principle of inferential justification, I noted that one need not take the principle to be the expression of a metaepistemological view at all. It could be accepted as a very general proposition of normative epistemology. In fact, though, I think that the principle is best understood as an implicit commitment to a certain analysis of inferential justification.

What is it to be inferentially justified in believing *P*? Well, if one looks at paradigm examples of appeals to evidence in support of belief, it is prima facie plausible to suggest that one's belief in some proposition *E* can justify one in believing another proposition *P* only when one's belief that *E* is itself justified and one has justification for thinking that *E* makes *P* probable. If asked why the conditional is true—indeed, necessarily true—it is tempting to suggest that it just describes that of which having evidence consists. As I said, one can make the claim initially plausible simply by looking at the ways in which it seems appropriate to *challenge* someone's claim to have good (epistemic) reasons for believing something. If I tell you that the world will end this century and offer as my evidence that there is an omnipotent God who has decided that he will destroy the world at midnight on January 31, 1999, you are presumably perfectly entitled to challenge the reasonability of my belief about the earth's extinction by challenging the reasonability of my belief about God. If, for example, I admit that I have no reason for thinking that there is a God with this desire, you can surely *for that reason* dismiss my claim to have inferential justification. Similarly, if I am talking to an astrologer who infers from the present alignment of planets that there will be prosperity this year, I am perfectly entitled to challenge the reasonability of this conclusion by challenging the reasonability of the astrologer's thinking that there is a connection between the two states of affairs. If the astrologer shrugs her shoulders and admits it is just a whimsical hunch that Jupiter's alignment with Mars might have something to do with economic prosperity here on earth, I can *for that reason* dismiss the astrologer's claim to have a justified belief about prosperity based on the position of planets relative to one another. I underscore "for that reason" in anticipation of the externalist's objection that I am guilty of level confusion.

The externalist will probably argue that when one raises questions about the astrologer's reasons for thinking that there is a connection between the alignment of planets and the affairs of humans, one is challenging the astrologer's reasons for thinking that she has a justified belief. One is not directly attempting to establish that the belief is unjustified. Because one expects, ceteris paribus, people to conform their beliefs to what they have reason to think is rational, one expects a person to abandon a belief when faced with the fact that they have no reason to think that it is rational. But on the face of it, this seems like a convoluted attempt to disguise the obvious. I am not merely telling the astrologer that she has no reason to believe that her belief is justified, I am telling the astrologer that her belief is unjustified, and a sufficient condition for her belief's being unjustified is that she does not have the slightest reason in the world for thinking that the positions of planets have anything to do with rising economies.

The second clause of the principle of inferential justification is rejected even by some philosophers who share at least some internalist intuitions. In "An Internalist Externalism" Alston offers a kind of compromise between access internalism and access externalism. Alston distinguishes the *ground* of our justification for believing *P* from the *adequacy* of that ground. A ground for *S* to believe *P* is some fact about *S* that makes likely the truth of *P*.[14] For *S* to be justified in believing *P*, according to Alston, *S* must have grounds for believing *P* and the grounds must be the kind of thing to which *S* has potential access. On the other hand, Alston thinks it would be far too strong to require that *S* be able to access the *adequacy* of these grounds, that is, the *connection* between the grounds and that for which they are grounds (p. 239).

I suspect that Alston's primary concern with the stronger requirement (our analogue of the second clause of the principle of inferential justification) is just the specter of skepticism. He is worried that for some commonplace beliefs most people (indeed, most philosophers) cannot figure out how the various candidates for justifying grounds do make probable the truth of the propositions for which they are alleged grounds. As I noted in chapter 2, the second clause of the principle of inferential justification is a powerful weapon in the skeptic's arsenal, and Alston is no doubt right to be worried. But why even require that to have justified belief one must have potential access to justifying grounds (though not to the fact that they are justifying grounds)? Alston's answer to this question is revealing.

Like many other externalists Alston makes the distinction alluded to earlier between *S*'s being justified in believing *P* and *S*'s *justifying* a belief that *P*, where the latter involves showing or arguing for the conclusion that one is justified in believing *P*. But though he insists that the concepts are distinct, Alston appears to argue that there is an intimate connection between them. We wouldn't have any *interest* in a concept of justified belief that is not compatible with our discovering the grounds of that justification (p. 236). But if Alston is right about this, it makes no sense to require access to grounds and not require access to their adequacy. To show that a belief is justified one would need to establish that some state *is* a justifying ground of that belief, and this would involve establishing the relevant connection between the ground and the truth of the proposition supported by that ground.

As I said earlier, I do not think that the principle of inferential justification does confuse conditions for being inferentially justified with conditions for being justified in claiming to be inferentially justified. We will reject our astrologer's beliefs as unjustified *for the reason* that the astrologer has no reason to believe that there is a probabilisitic connection between astrological evidence and astrological predictions. It does *seem* that we commonly grant inferential justification in other cases where one has no access to the inferential connection, but this may reflect only that we take for granted the existence of such access without thinking reflectively (philosophically) about the relevant inferences. I daresay if we were part of a community of committed and unreflective astrologers, we might be quite satisfied that we had astrologically justified beliefs without ever wondering how we satisfy the second clause of the principle of inferential justification. But if upon philosophical prodding we suddenly realized that we had no reason to think that there were any legitimate astrological inferences, we would (or at least should) abandon *for that reason* any claim to have astrologically justified belief.

An anonymous referee once suggested to me that the second clause of the principle of inferential justification should be replaced with one that makes no reference to probability. Specifically, one might claim that to be justified in believing *P* on the basis of *E*, one must be (1) justified in believing *E* and (2) justified in believing that if *E* then *P*. But one must immediately inquire as to the interpretation of the conditional, if *E* then *P*. One thing seems obvious. If one is trying to understand how one could be justified in believing *P* *through* or *by* justifiably believing the conjunction of *E* and if *E* then *P*, one should not interpret the conditional as a contingent, truth-functionally complex conditional of material implication that is not made true by a *connection* between *E* and *P*. *E* materially implies *P* just means either *not-E* or *P*. Now when one is justified in believing *E* (that is, when one satisfies the first clause of the principle of inferential justification), then the only way to be justified in believing *not-E* or *P* (in the absence of justifiably believing that there is a connection between *E* and *P*) is to justifiably believe *P*. Our revised principle, then, would tell us that to be justified in believing *P* on the basis of *E*, one must be justified in believing *E* and justified in believing *P*. But if we already had justification for believing *P*, one would not need to infer *P* from *E*![15]

The only conditional that one could plausibly appeal to in a principle of inferential justification is one that asserts a *connection* between *E* and *P*. But what kind of connection should we view the conditional as asserting? Entailment is one candidate, but we can surely justifiably believe *P* on the basis of *E* even when *E* does not entail the truth of *P*. What we seem to need is the kind of connection that holds between the premises and conclusion of a good argument. When a deductively invalid argument is good, the premises must make probable the conclusion. To be justified in believing *P* on the basis of *E*, one must be justified in believing that the inference from *E* to *P* is legitimate. To justifiably believe that an inference from *E* to *P* is legitimate just *is* to justifiably believe that *E* makes probable *P* (where *E*'s entailing *P* can be thought of as the upper limit of making probable, that is, the case where the probability of *P* given *E* is 1). But now we are back to the earlier version of the principle of inferential justification. To be justified in believing *P* on the basis of *E*, one must be justified in believing that *E* makes probable *P* (that is, that the inference from *E* to *P* is legitimate).

But are there not other connections between *E* and *P* that we could justifiably believe obtain and that would give us justification for believing *P* provided that we have justification for believing *E*? Suppose, for example, we know that there is a causal connection between *E* and *P.* If we know that *E* is the cause of *P* and we know that *E*, can we not know on that basis that *P*? The answer is yes, but only because this will be a case of deductively inferring *P* from *F* where *F* = the conjunction of *E* and *E* causes *P.* That conjunction *entails P* (on at least most conceptions of causality), and if we are aware of the entailment relationship, we will have inferential justification for believing *P* by satisfying the principle of inferential justification as it applies to the inference from *F* to *P.* But as epistemologists we will want to know what the evidence is for thinking that *E* causes *P.* Call that evidence *E1.* The philosophically interesting question will concern the relationship between *E1* and the proposition that *E* causes *P.* And on the suggestion I have made, to justifiably believe that there is a causal connection on the basis of *E1* one would have to know that the reasoning from *E1* to the proposition asserting causal connection is legitimate, that is, that *E1* makes probable the existence of a causal connection.

A Conceptual Regress Argument for Foundationalism

If we are going to *analyze* inferential justification employing our concept of noninferential justification and using the principle of inferential justification as our guide, what will our analysis look like? To be successful our analysis will eventually need to eliminate the use of the term "justified." *S* is justified in believing *P* on the basis of *E* only if *S* is justified in believing *E* and justified in believing that *E* makes probable *P.* But what is the analysis of justification? When we considered traditional arguments for foundationalism earlier, we emphasized the way in which acceptance of the principle of inferential justification might seem to give rise to a vicious *epistemic* regress. The traditional foundationalist, anxious to avoid skepticism, worries that without foundations, having a justified belief would entail having an infinite number of different justified beliefs. But if we are building the principle of inferential justification into an analysis of the very concept of justification, we have a more fundamental vicious *conceptual* regress to end. We need the concept of a noninferentially justified belief not only to end the epistemic regress but to provide a conceptual building block upon which we can understand all other sorts of justification. I would argue that the concept of noninferential justification is needed (whether one is a skeptic or not) in order to *understand* other sorts of justification, much the same way that the concept of something being good in itself is needed in order to understand other ways in which things can be good. Consider the analogy.

Notice that one could argue for the concept of intrinsic goodness as a way of avoiding an epistemic regress. Suppose we crudely define being good as a means in terms of producing something that is good. A philosopher could argue that not everything is good as a means, for if everything were good as a means, a finite mind could never know, or even be justified in believing, that something is good. To know that *X* is good I would have to know that it leads to something else *Y* that is good, but if everything is good only as a means, then to know that *Y* is good I would have to know that it leads to something else *Z* which is good, and so on ad infinitum. If one then includes as a premise a rejection of radical skepticism with respect to knowledge of what is good, one has an argument for the view that some of the things we know to be good we know to be intrinsically good. But one could just as easily leave the epistemology out of it. It seems to me that the view that there is only instrumental goodness is literally unintelligible. To think that something *X* is good if all goodness is instrumental is to think that *X* leads to a *Y* that is good by virtue of leading to a *Z* that is good, by virtue of. . . , and so on ad infinitum. But this is a vicious conceptual regress. The thought that *X* is good, on the view that all goodness is instrumental, is a thought that one could not in principle complete. The thought that a belief is justified, on the view that all justification is inferential, is similarly, the foundationalist might argue, a thought that one could never complete.

Just as one terminates a conceptual regress involving goodness with the concept of something being intrinsically good, so one terminates a conceptual regress involving justification with the concept of a noninferentially justified belief. On the internalist account of noninferential justification just sketched, the concept of a noninferentially justified belief is ultimately explicated in terms of the sui generis relation of acquaintance. As an initial attempt, then, we might try to define justification in terms of noninferential justification as follows:

S is justified in believing *P* when (1) *S* is noninferentially justified in believing *P* OR (2) there is some proposition *E* such that *S* is noninferentially justified in believing *E* and that *E* makes probable *P* OR if *S* is not noninferentially justified in believing *E*, there is some other proposition *E1* such that *S* is noninferentially justified in believing *E1* and that *E1* makes probable *E* OR if *S* is not. . . , AND if *S* is not noninferentially justified in believing that *E* makes probable *P*, then there is some proposition *F* such that *S* is noninferentially justified in believing *F* and that *F* makes probable that *E* makes probable *P* OR if *S* is not noninferentially justified in believing *F* . . .

This has the familiar form of a recursive analysis of justification that is now so familiar to us through Goldman's attempt to provide a recursive reliabilist analysis of justification. The base clause is *S*'s being noninferentially justified in believing *P*. If the base clause is not satisfied, then *S*'s justification must be traceable to noninferentially justified beliefs. But because as inferential internalists we are accepting *both* clauses of the principle of inferential justification, once inference is involved there must always be at least two noninferentially justified beliefs. One provides one with premises, as it were. The other provides one with the appropriate evidential connection between premises and conclusion.

The above analysis is unsatisfactory as it stands. There are two problems, one dealing with defeaters of justification, the other dealing with loss of probability through repeated inferences. Consider the first. Intuitively, I may have an impeccable, foundationally terminated chain of reasoning, but the resulting belief is unjustified because there is other information and other valid reasoning available to me which, had it been used in addition to the reasoning I did use, would have led me to the opposite conclusion. Second, it is a well-known feature of probability relations that they are not transitive. From the fact that *P* makes probable *Q* and *Q* makes probable *R* it does not follow that *P* makes probable *R*. And this is so not only because the possibility of error "adds up" with each inference. In fact, one can give examples where P makes probable Q and Q makes probable R, but P makes probable not-R. That $n = 2$ entails (the upper limit of making probable) that *n* is prime, and that *n* is prime makes likely that *n* is odd. But that $n = 2$ does not make likely that *n* is odd—indeed, it entails that it is not odd.

These problems also afflict other *externalist* analyses of justification, and I suggest that we postpone an attempt to solve them until the next chapter when we see how they arise in connection with reliabilism. The sort of solution we propose for our traditional recursive analysis of justification may also be available to the externalist.

Before turning to externalist versions of foundationalism, it might be useful to make a distinction here, perhaps overdue, between the existence of justification for a belief and a belief's actually being justified. Suppose that *S* is noninferentially justified in believing both that *E* and that *E* makes probable *P* (where *S* has no other evidence bearing on the truth of P), but that S does not believe P. In this situation I say that S has justification for believing P but does not have a justified belief that P (for S does not even believe that P). Suppose that S is noninferentially justified in believing E and that E makes probable P (where no other evidence bears on the truth of P) and S does believe P, but what causally sustains S's belief is something other than the good epistemic reasons S possesses. Although it is a matter of some controversy, many philosophers would again refuse to count S's *belief* as justified. Even if S has justification for believing P, the belief will not be justified unless the justification S has causally sustains the belief.[16] The existence of a nomological connection between S's justification and a resulting belief, though necessary, is probably not sufficient for the resulting belief's being justified. As Audi (1993, chapter 8) argues in a sophisticated discussion of different concepts of "basing" and "inference," there are all kinds of puzzles arising from the possibility of deviant causal chains. Thus, if I justifiably believe E and justifiably believe that E makes probable P, and this somehow causes my psychiatrist to use hypnosis to induce my believing P, many philosophers would still refuse to regard my belief as justified. What one probably needs is the concept of a nondeviant causal chain,[17] but the problem of defining clearly "relevant" causal connection so as to make the concept immune to counterexamples has plagued not only causal theorists of "basing" relations but causal theorists of perception, reference, and intentionality (to give just a few examples).

Although it is not at all clear that ordinary discourse will always insist on the distinction made here between beliefs for which one has justification and beliefs that are causally sustained in the appropriate way by such justification, it does seem to be a useful distinction nevertheless, and I will continue to mark it using the locutions as indicated earlier. The expression "S has justification for believing P" will be used in such a way that it implies nothing about the causal role played by that justification in sustaining belief. The expression "S's belief that P is justified" will be taken to imply both that S has justification and that S's justification is playing the appropriate causal role in sustaining S's belief. It seems obvious that in the context of normative epistemology, philosophers qua philosophers should be interested primarily, or perhaps even exclusively, in questions as they relate to having justification. Once one settles whether or not one has justification for believing some proposition, the question of whether the belief is justified becomes an empirical question concerning causal connections, an empirical question that goes beyond the scope of philosophical investigation.

Notes

1. See Luper-Foy 1985 for the argument that knowledge is "external."
2. BonJour (1985) is perhaps the paradigm of an internalist who understands the view this way.
3. It seems to me that this is the strongest form of access that Chisholm would recognize as necessary for justification. See Chisholm 1989, p. 76.
4. I discuss Alston's view later in this chapter.
5. In one of the most sophisticated attempts to combine the insights of contemporary externalism, traditional foundationalism, and the coherence theory of justification, Ernest Sosa makes a move very similar to this. While he is a fairly straightforward externalist with respect to what he calls *animal* knowledge (requiring no access requirements for the person who has such knowledge), he feels the force of requiring some sort of access to the conditions constituting animal knowledge if a person is to have *reflective* knowledge (the kind of knowledge that a philosopher will seek). For reflective knowledge Sosa does require a kind of access to the source of a belief being reliable (apt). Unlike the externalist I am imagining here, however, it seems to me that Sosa turns to coherence (albeit coherence externally understood) as the way in which one achieves access to the fact that one's beliefs are appropriately (virtuously) produced. See Sosa 1991: the point is made throughout a number of articles contained in the collection.
6. Goldman's attempt to define the reliability of a process in terms of whether it would produce mostly true beliefs in *normal* worlds (Goldman 1986) probably abandons nomological conceptions of reliability. But Goldman has now abandoned that idea.
7. Goldman 1979, 1986, 1988; Nozick 1981; Dretske 1969, 1981; and Armstrong 1968.
8. Descartes 1960; Hume 1888; Chisholm 1957, 1989; Price 1950; and Russell 1926, 1959.
9. Ayer (1956, 19) presents this argument.
10. See BonJour 1985, pp. 58–78.
11. The argument is given in Sellars 1963, pp. 131–32 and also in BonJour 1985, chap. 4.
12. See Goldman 1979, p. 1.
13. In Chisholm 1989 the primitive conceptual building block for the analysis of other epistemic concepts is *S* being more justified in believing one proposition than another.
14. Alston is very cautious in saying what this probability connection is, but it is clear that it involves an "objective" connection of some kind. I discuss probability relations further in chapter 7.
15. It will, of course, be true that someone who satisfies the two clauses of this revised principle of inferential justification will be justified in believing *P*. The question is whether we will have offered an illuminating account of how one gets justification *through* inference.

16. For an excellent discussion of the distinction, see Audi 1993, especially chap. 8. For a view that downplays the importance of causal requirements, see Foley 1987, chap. 4.

17. See again Audi 1993, pp. 257–62, for a more detailed discussion of kinds of problematic causal intermediaries and how one might handle them in analyzing what it is for someone to believe something *for* a reason.

Chapter Four

Externalist Versions of Foundationalism

In the last chapter we looked at traditional versions of foundationalism often associated with internalism. I developed in more detail the version of traditional foundationalism that I take to be most plausible. That foundationalism ends the epistemic and conceptual regress of justification with the concept of direct acquaintance, a concept that is sui generis, that cannot be reduced to any more fundamental concepts. Having distinguished a number of different theses often associated with the technical distinction between internalist and externalist epistemologies, I made clear the senses in which an acquaintance theory of noninferential justification is and is not internalist in structure. The specific acquaintance theory I defended is *consistent* with "internal state" internalism, provided that we identify internal states with nonrelational properties of the mind and those relations whose relata involve only the mind and its nonrelational properties. I cautioned, however, that one can be an acquaintance theorist who takes the relata of acquaintance to include entities external to the mind, and it is difficult to include such a theory as a species of "internal state" internalism. I then argued that an acquaintance theorist (and any other metaepistemologist) would do well to stay away from strong access internalism, a view that simply invites a vicious regress. Although traditional foundationalists can often accommodate versions of weak access internalism, I argued that it might be a mistake to view one's commitment to weak access internalism as lying at the heart of the internalist/externalist debate. In most of the senses in which an internalist can incorporate a weak access requirement, so can an externalist without leaving the framework of externalism. Finally, I argued that acquaintance theories and other paradigm traditional foundationalisms might best be thought of as

essentially different from externalism by virtue of their refusal to reduce fundamental epistemic concepts to other concepts, particularly the
nomological concepts on which the externalist so heavily relies. In
other words, the essence of internalism might be its refusal to "naturalize" epistemology. I also argued that there is a genuine defining distinction between internalists and externalists in terms of their willingness
to incorporate both clauses of the principle of inferential justification
into an analysis of justification. Paradigm internalists will insist that
justification that is owed to inference requires that the believer have
epistemic access to the legitimacy of the inference. Paradigm externalists deny this. I suggested we call proponents of these respective positions inferential internalists and inferential externalists.

I now examine the structure of some of the paradigm externalist
analyses of justification in light of the conclusions we have tentatively
reached about the heart of the internalism/externalism controversy. Although I examine these views critically, my primary interest is to pin-
point the similarities and essential differences between internalist and
externalist versions of foundationalism, and to pave the way for our
subsequent discussion of the implications of these views for the way in
which we should approach the skeptical challenge. Because it is still
one of the most influential, well-developed, and sophisticated versions
of externalism, I take Goldman's reliabilism as a paradigm of externalism. Through an examination of it I try to reach certain general conclusions about the nature of externalism. After discussing Goldman I comment briefly on other versions of externalism, specifically Nozick's
well-known analysis, in order to ensure that the generalizations I reach
by looking at reliabilism apply to other paradigmatic externalist analyses of epistemic concepts.

Goldman's Reliabilism

If a philosopher's importance is to be measured by the effect that
philosopher has on the field, Goldman's contribution to epistemology
is enormous. Although the naturalizing of epistemology can perhaps be
traced to more than one source,[1] it is unquestionable that it is Goldman's work that has captured the imagination of many contemporary
epistemologists and has threatened a metaepistemological revolution of
Copernican scope. If Goldman is right about the way in which epistemologists should understand fundamental epistemic terms, one should
view much of the history of epistemology as seriously flawed. The phil-

osophical response to the threat of skepticism, indeed the willingness to treat skepticism seriously as a philosophical issue, depends crucially on the plausibility of Goldman's metaepistemology. Goldman's views about justification have changed dramatically from the now classic paper "What is Justified Belief?" through his book *Epistemology and Cognition*, and yet again in more recent papers. Let us begin with the early paper that was so influential.

Before examining the details of Goldman's analysis, it is worth emphasizing again that Goldman did not pretend to be offering a *definition* of what he takes to be the *normative* concept of justified belief. I take it that the naturalistic analysis he offers of justification is intended to capture the natural properties upon which the normative concept of justification *supervenes*. The supervenience in question is *not* merely nomological, however. One can test the adequacy of the analysis much the way one tests the adequacy of a more traditional reductive analysis. The conditions that constitute the analysans must be necessary and sufficient for the analysandum, and to evaluate the relevant claims of necessity and sufficiency we can let our imaginations roam over all possible worlds. Partly because of the nature of Goldman's analysis, it will also be easiest to focus in subsequent discussion on the conditions for a belief's being justified as opposed to the conditions that make it true that there is justification for a belief (see the distinction made at the end of the last chapter).

The fundamental idea behind Goldman's reliabilism is straightforward enough. When a belief is justified it has a virtue. There is something good about it. From the epistemic perspective, virtue has to do with *truth*. The reason epistemologists want epistemically justified beliefs, it is presumed, is that having justified beliefs has *something* to do with having true beliefs. At the same time, we must understand justification in such a way that we allow the possibility of justified false belief. How do we establish a connection between justification and truth without making it impossible to have a justified false belief? The answer is to focus on the processes that produce beliefs. The beliefs that are "good" from the epistemic perspective are those that are produced by reliable belief-forming mechanisms. Reliable belief-forming mechanisms are those that usually get you true beliefs, and since reliability is not an all-or-nothing matter, a false belief can be produced by a reliable belief-forming mechanism.

This sort of sketch of the reliabilist's conception of justification is almost always misleading. For one thing it is not true that reliable belief-forming mechanisms usually get you true beliefs. In the case of belief-forming mechanisms that take as their input beliefs, the "output" is only going to be as good as the "input." Because Goldman emphasizes the importance of distinguishing between belief-forming mechanisms that take as their input beliefs and belief-forming mechanisms that take as their input other stimuli, the analysis of justification he offers has the familiar structure of foundationalism. Just as did the traditional foundationalist, Goldman distinguishes beliefs that owe their justification to other justified beliefs from beliefs that do not owe their justification to the having of other justified beliefs, and I assume he would have no objection to using our terminology, "inferentially and noninferentially justified beliefs," to mark the distinction. The subsequent recursive analysis of justification he offers will use in its base clause the concept of a noninferentially justified belief.

What is a noninferentially justified belief for Goldman? *Initially*, the idea is relatively simple. Some belief-forming mechanisms take as their "input" stimuli that do not include beliefs. He calls these *belief-independent* processes and characterizes the kind of reliability they can have as *unconditional*. The crudest way of understanding their reliability would be in a straightforward statistical sense. A belief-independent process is unconditionally reliable when most of the beliefs it actually produces are true. As we shall see, Goldman, and reliabilists generally, will not accept this sort of statistical understanding of reliability, and as they move away from it, it will be interesting to see how successful the view is in accomplishing its initial goal of securing a connection of some kind between having justified beliefs and having (mostly) true beliefs.

The concept of a belief resulting from a belief-independent, unconditionally reliable process parallels in many respects the older concept of a noninferentially justified belief that owes its justification to direct acquaintance with facts. But it is also importantly different. There is no temptation on the reliabilist view to associate noninferential justification with infallibility. Belief-independent, unconditionally reliable processes might be only 70 percent reliable and the beliefs produced will still be noninferentially justified. It is interesting to note, however, that the reliabilist can easily recognize a subclass of beliefs that are noninferential and infallibly justified. These will be the beliefs, if there are any, that result from belief-independent processes that are 100 percent reliable. When I have this sort of justification for believing *P*, my justification will entail the truth of *P*. Because noninferential justification on the reliabilist model need not be very strong justification, there are all kinds of ways in which the reliabilist can attempt to escape the traditional skeptical argument that seeks to establish inferential justification as the only sort of justification for believing propositions about the past, the external world, and the future, and we will explore these options in considerable detail when we look at the appropriate externalist response to skepticism.

Just as on the traditional foundationalist view noninferentially justified beliefs can be used to justify still other beliefs, so on Goldman's view the beliefs that result from belief-independent, unconditionally reliable processes can in turn be taken as input, processed, and result in additional beliefs. These additional beliefs will be justified provided that the *belief-dependent* processes that produced them are *conditionally* reliable. The crudest (certain to be eventually abandoned) statistical characterization of conditional reliability is this: A belief-forming process is conditionally reliable when its output beliefs are usually true when its input beliefs are true. Inferentially justified beliefs, then, are those that result from belief-dependent, conditionally reliable processes whose input beliefs are themselves *justified*. The reliabilist traces all justification back to noninferential justification and offers a recursive analysis of justified belief. Unlike our inferential internalist, the reliabilist will *not* build into the recursive analysis any reference to justified belief about the legitimacy of the inferential processes. The reliabilist is an inferential externalist. Even when the process resulting in a belief takes as its input other beliefs, the believer does not need to have any reason to suppose that the belief-forming process is reliable in order to have a justified belief. So the crudest recursive analysis of justification Goldman considers looks like this: A belief is justified when it results from a belief-independent process that is unconditionally reliable OR when it results from a conditionally reliable belief-dependent process whose input beliefs result from belief-independent processes that are unconditionally reliable OR when it results from a conditionally reliable belief-dependent process whose input beliefs result from belief-dependent conditionally reliable processes whose input beliefs result from belief-independent unconditionally reliable processes OR . . . . The analysis, in effect, takes the form of an infinitely complex disjunction of sufficient conditions for a belief's being justified. The mind can grasp the infinite complexity because it understands and can project the pattern.

Internal Difficulties for Reliabilism and Alternatives to Recursive Versions of both Reliabilist and Traditional Foundationalisms

We can divide the problems this sort of analysis faces into fundamental objections to the very idea of reliabilism and internal difficulties that arise even if we presuppose the general plausibility of the approach. Let us consider the latter first. Goldman himself recognized that he must qualify the analysis to deal with a problem alluded to in our earlier discussion of the acquaintance recursive analysis of justification. Intuitively, it seems that we can have an impeccable chain of inferences leading from a noninferentially justified belief to a belief that *P*, which is nevertheless unjustified because of additional data which, though not used, were available and would have led by equally legitimate reasoning to *not-P*. Goldman gives the example of a person whose beliefs about the past are produced by reliable ''memory processes'' but who has ample evidence (misleading, as it turns out) indicating that these processes are unreliable. He was told, let us suppose, by all sorts of people whom he should have taken to be reliable authorities that he is a victim of massive, illusory, drug-induced memory experiences. Notice that the problem seems to affect even beliefs that satisfy the initial *base* clause sufficient condition for justification. Indeed, beliefs about the past produced by memory might very well be beliefs that are noninferentially justified on Goldman's view—they may be beliefs that result from belief-independent, conditionally reliable processes. In order to resolve this problem, Goldman suggests the following revision of his base clause (the clause in terms of which other justification is ultimately to be understood):

> If *S*'s belief in *p* at *t* results from a reliable cognitive process, and there is no reliable or conditionally reliable process available to *S* which, had it been used by *S* in addition to the process actually used, would have resulted in *S*'s not believing *p* at *t*, then *S*'s belief in *p* at *t* is justified.[2]

A number of questions arise concerning this modified base clause, some of which Goldman discusses. In particular, it will be difficult to indicate what makes a process that is not operating in *S*, one that is nevertheless available to *S*. Once again the disguised modal operator present in the concept of availability admits of numerous interpretations. The logical possibility of using a process is too broad, and so perhaps is lawful possibility. The possibility is probably best understood in terms of that more common concept that finds expression in ordinary discourse but is difficult to define precisely. Roughly, a process is available to one if a suitably restricted description of one's present characteristics does not lawfully preclude the use of the process. I say suitably restricted, because if the world is deterministic, a complete description of the antecedent states of the subject might well lawfully preclude the use of any process other than the one actually used.[3]

My primary concern here, however, is not with the concept of availability, but rather with a necessary part of the revised base clause that seems to have been left out. When he introduces the revision, Goldman tells us that he will "'omit certain details in the interest of clarity'" (p. 20). One omitted detail that should certainly concern us is the implicit reference to *justified* belief contained in the reference to available conditionally reliable processes. The availability of a conditionally reliable process that would lead S to not believe p would be relevant, I assume, only if one had *justified* beliefs that the conditionally reliable process could use as input. The base clause, stated more completely and clearly, becomes:

> If S's belief in p at t results from a reliable cognitive process, and there is no belief-independent, unconditionally reliable process which had it been used by S would have resulted in his not believing p, and there is no belief-dependent process which is conditionally reliable that could have been used by S to process certain *justified* beliefs so as to result in him not believing p at t, then S's belief in p at t is justified.

But as stated, such a base clause manifestly fails to achieve the purpose of a base clause. A metaepistemological analysis of justification offered with such a base clause fails to end a vicious *conceptual* regress involved in trying to understand the concept of justified belief.

There appears to be a quite general dilemma for reliabilists here. If they accept the conceptual regress argument for the necessity of introducing a concept of noninferential justification, then either they return to the crude view according to which the mere fact that a belief is produced by an unconditionally reliable process makes it justified and face obvious counterexamples, or they add the kind of qualifications that Goldman does and run afoul of the conceptual regress argument. Moreover, I think there are more general morals to be drawn. In order to end a conceptual regress involved in trying to understand the concept of justified belief, one seems to need an account of noninferential justification that does not make reference to the availability of various sorts of *inferential* justification. If one is an acquaintance theorist, for example, one cannot resolve the conceptual regress problem of understanding justification by understanding a noninferentially justified belief as one held while one is acquainted with the relevant facts *and* one has no other justified beliefs from which one can infer a conflicting conclusion. One simply cannot make use of a clause referring to justification in the base clause of a recursive analysis of justification. The whole idea behind understanding justification ultimately by reference to noninferential justification is based on the supposition that understanding the truth conditions for inferential justification is parasitic upon understanding the truth conditions for noninferential justification, and not vice versa.

You will recall that the more traditional recursive analysis of justification involving the base clauses dealing with acquaintance itself faces a similar difficulty to the one just discussed in connection with Goldman's analysis. Like Goldman, we acknowledge that having a belief with impeccable justificatory ancestry is not enough to have a justified belief. One must take into account other evidence that one could and should have used in addition to the evidence from which one did infer one's conclusion. I suggest that there is a way for both the reliabilist and the more traditional foundationalist to avoid the problem discussed above, and that is to employ counterfactuals dealing with the hypothetical situation in which all belief-forming processes (in the externalist's language) and all relevant inferences (in the terminology of more traditional foundationalists) are employed.

Let me illustrate the suggestions first with respect to the reliabilist. And let us begin by invoking the distinction made at the end of the last chapter between someone's *having* justification for believing something and that belief's *being* justified. I first attempt to offer on behalf of the reliabilist an analysis of what it is for S to have justification for believing P (whether or not that justification is causally producing or sustaining the belief in question). We can begin with the "pure" base clause dealing with beliefs that result from unconditionally reliable belief-independent process. With such a clause we have made no use of the concept of justification we are trying to analyze. Goldman wants to use this clause or a revised version of it to find a sufficient condition for justification that can be used in a recursive analysis of justification. Instead, he could simply describe a hypothetical situation in which all of the reliable belief-independent processes available to S are used simultaneously on all of the data presently available to S while all of the reliable belief-dependent processes available to S are used on all of the data generated by both reliable belief-independent and belief-dependent processes. He could further add to a description of the hypothetical situation the stipulation that there are no other processes operating that are causally relevant to the occurrence of belief. If in this hypothetical situation S would believe P, then the reliabilist could hold that S has justification for believing P. If S would not believe P, then S does not have justification for believing P. We would avoid the aforementioned difficulties with a standard recursive analysis and have a much more straightforward version of reliabilism as a bonus:

S has justification for believing *P* at *t* if the totality of reliable belief-forming processes available to *S* (belief-independent and belief-dependent) operating on the totality of data available to *S* (stimuli in the case of belief-independent processes, beliefs in the case of belief-dependent processes) would result in *S*'s believing *P* at *t* (when no other processes capable of producing beliefs are operative).

As I noted earlier, most philosophers will not take having justification for believing *P* when one has the belief that *P* to be a sufficient condition for the belief's being justified. They hold that in some sense or other justified beliefs must be held *as a result* of the justification one possesses. If I believe in my children's honesty as a result of blind faith, then even if I have ample evidence to support the conclusion that they are honest, I will not have a justified belief. If we accept this conclusion, then to move from the above reliabilist account of having justification for believing *P* to an account of having a justified belief that *P*, we must add some condition referring to the causal source and sustenance of belief. It will probably be too strong to require that the having of justification is itself causally explaining the belief. Rather, it seems enough that the *actual* processes resulting in the belief be legitimate while the person in question *possesses* justification in the sense defined above. Thus we could offer the following account of *S*'s being justified in believing *P*:

> *S*'s belief that *P* at *t* is justified if *S* has justification for believing *P* and the belief-forming processes actually responsible for *S*'s belief that *P* are unconditionally or conditionally reliable.

As presented, this account attempts to state only a sufficient condition for a belief's being justified. Intuitions may vary, but most will be inclined to allow that justified beliefs can be causally supported by at least some defective belief-forming processes provided that they are not *essential* to the formation of the belief, that is, they could be absent and the belief would still occur as a result of legitimate reasoning.

It is an interesting feature of philosophical intuitions about these matters that we tend to think that otherwise legitimate reasoning can fail to yield justified belief because of the availability of reasoning that would lead to an opposite conclusion, while otherwise illegitimate reasoning cannot succeed in producing a justified belief when there is available legitimate reasoning that would support the conclusion. Symmetry might seem to suggest that what can be taken away by available reasoning can likewise be restored. Perhaps our intuitions reflect primarily the

idea that if what is causally operative in producing belief involves defects from the epistemic perspective, then our evaluation of the belief
so formed should reflect that fact and we should mark it as unjustified.
Given the above reliabilist analyses of S's having justification for
believing a proposition, and S's belief in a proposition being justified,
the concepts of epistemic justification would still be parasitic upon the
fundamental concept of a belief-independent process. Unlike standard
recursive analyses, we would no longer be viewing the operation of any
single belief-forming process as sufficient for the having of a justified
belief.

The approach I would take to the more traditional foundationalist
analysis is somewhat different. I am prepared to claim that the three
acts of acquaintance I discussed are sufficient for a belief's being justified because they are *incorrigible*. Let us say that S's belief that P at t
is incorrigible if the justification S has for believing P at t could not be
destroyed *at t*. Incorrigibility is, of course, different from infallibility.
From the fact that it is impossible for me to possess any better or more
complete evidence than the evidence I have, it does not follow that that
evidence precludes the possibility of error. But even if the three acts of
acquaintance constitute having noninferential justification, and even if
we have a chain of reasoning that goes all the way back to a noninferentially justified belief where we have noninferential justification for accepting the legitimacy of each inference in the chain, the resulting belief
might not be justified. It might not be justified because there might
be a perfectly legitimate chain of reasoning going back to a different
foundation which results in the opposite conclusion. If we again begin
by attempting to characterize what it is for a person to *have* justification
for believing a proposition, we might do so by talking about what the
resulting beliefs would be if all available inferences were used from all
of the available noninferentially justified beliefs.

Given the more traditional foundationalism, we could restrict the relevant inferences to *primary* inferences. Let us say that an inferential
connection between E and P is primary for S if S is noninferentially
justified in believing that E makes P probable. As we ordinarily talk we
say that we can infer that a solution is acid from the fact that the litmus
paper turns red, or we can infer that there will be a storm from the
presence of very dark clouds. But the availability of these "inferences"
is clearly parasitic upon the availability of more complex arguments.
The question of how to distinguish primary from secondary inferences
will of necessity be a matter of much controversy. Obviously, given
the foundationalism I prefer, it will be possible to identify the primary

inferences as those which we can be noninferentially justified in accepting. Many philosophers would also feel comfortable making the distinction in terms of necessary evidential connections versus contingent evidential connections. That the premises of an inductive argument make probable its conclusion, some philosophers would argue, is a necessary truth. That acidic solutions turn litmus paper red is a contingent fact. Neither of these approaches would be accepted by the reliabilist. The legitimacy of any nondeductive belief-forming process will always be a contingent matter for the reliabilist. The reliabilist can nevertheless still make the distinction between rules of inference that are derived from the use of other rules and rules of inference that are not.[4]

The reason we need refer only to the operation of primary inferences in our analysis of inferential justification is that if all of the primary inferences are operating on all of the available noninferentially justified premises, we will automatically include all of the relevant secondary probabilistic connections. Given the truth of foundationalism, if we have justification for accepting the secondary epistemic principles, that justification can be traced to noninferentially justified beliefs and the legitimate primary inferences that can be made from those foundational truths. So our revised analysis of having justification for believing *P*, given the traditional foundationalism discussed in chapter 3, will read as follows:

S has justification for believing *P* at *t* when either *S* is noninferentially justified in believing *P* OR if *S* were to employ all of the inferences directly available to him on all of the propositions in which *S* has noninferentially justified beliefs as well as all of the beliefs in turn generated by legitimate inference, then *S* would believe *P* (provided that no other processes causally effective in producing or preventing belief are operative). An inference from *E* to *P* is directly available to *S* only if *S* is noninferentially justified in believing that *E* makes *P* probable.

It should be noted that if one wanted to embrace the central idea behind a direct acquaintance analysis of noninferential justification but also wanted to avoid the claim that acquaintance can provide one with incorrigible justification, one could easily modify this account to make it look more like the analysis I suggested for the reliabilist. The analysis of justification would omit the first disjunct and the conditional would simply talk about the results of employing all directly available inferences on propositions that have prima facie justification provided by acts of acquaintance.

For a belief to be actually justified, we might again want to require more than that the believer have justification for holding the belief. We might want to require in addition that there be some legitimate chain of reasoning leading from noninferentially justified beliefs to the conclusion in question, the existence of which is causally sustaining the belief. If the belief is causally overdetermined and is causally supported as well by illegitimate reasoning, that illegitimate reasoning must not be causally necessary for the belief to occur, that is, the belief in question would occur even in the absence of that faulty reasoning.

There may be an added bonus to the above alternatives to both reliabilist and traditional foundationalist recursive analyses of justification. The other problem alluded to in chapter 3 concerned the loss of probability through repeated legitimate nondeductive inferences. The possibility of error in each step of a long chain of reasoning might be slight, but if the chain is long enough the possibility of error somewhere between the beginning and the end might be huge. If you are as bad at addition as I am, you will recall that when adding a very long column of numbers you make errors more often than not. The chance of making an error at each step in the addition is small but the chance of error increases dramatically as the addition problem gets longer. This is a problem for both reliabilists and traditional foundationalists. Goldman recognizes the problem in a footnote[5] but offers no solution. But without a solution his recursive analysis of justification is simply implausible. If there are a large number of belief-dependent processes used in reaching a conclusion, then even if they are all reliable the conclusion might be highly improbable due to the mounting chance of error. The revised reliabilism suggested earlier might seem to take care of this problem because, since we can figure out all this stuff about mounting chances of error, the availability of this conclusion to be processed with all our other information should enable us to make the necessary corrections in our belief system. If we were still to end up believing *P* after using *all* of the relevant reliable belief-forming processes *including* those that inform us with respect to the mounting possibility of error, then presumably *P* would still be justified. And if this solution works for the reliabilist, it will, of course, also work for the revised, more traditional foundationalist analysis of justification.

Although promising, I am not sure in the end that this solution to loss of probability through long chains of reasoning is adequate. In the preceding paragraph I suggested that the relevant calculations concerning loss of probability are *available* to us. But earlier we noted that we must be careful how we construe availability. The logical possibility of use and even the lawful possibility of use seem too broad. The relevant possibility implicitly invoked in this concept of availability seems to require that we take into account not just what is lawfully possible, but what is lawfully possible given the current state of the subject. And to put it crudely, it seems almost certain that there are many people who are simply too stupid to realize that chains of reasoning unproblematic at each step involve an unacceptably high probability of error by the time the chain is complete. For these people, the relevant inference concerning this mounting probability of error will *not* be available. Is there any other way of understanding justification that both the reliabilist and the more traditional foundationalist might invoke to resolve this problem? I think there is but it involves controversial assumptions for both reliabilists and traditional foundationalists. First consider reliabilism.

The reliabilist talks about belief-independent and belief-dependent processes, which can be used in chains of reasoning. Each of these processes has a certain unconditional or conditional reliability. But is there any reason why one cannot talk about the reliability of the single large process that is constituted by the use of a number of other processes?[6] Suppose, for example, that I employ four reliable processes *P1*, *P2*, *P3*, and *P4* to arrive at the belief that *Q*. *P1* is a reliable belief-independent process that takes input *I* and churns out beliefs that are processed by *P2*, whose output beliefs are processed by *P3*, whose output beliefs are processed by *P4*, resulting in the belief that *Q*. Each of the processes is individually reliable, but why can we not talk about the reliability of the complex process we can describe as taking input by *P1* and processing the results through *P2*, *P3*, and *P4*? That complex process might still be generally reliable or it might be generally unreliable. The difficulty that such a solution seems to face is that it runs the risk of making processes overly specific. For example, the reliability of the whole process will no doubt depend in part on how *many* beliefs are outputs and inputs of the respective constitutive processes, and this will vary from case to case. We can resolve this problem by making our description of the whole process very specific, for example, the complex process consisting of *P1* producing eight beliefs that are processed by *P2* which turns out four beliefs, and so on. The danger of making the description of a process too specific is that there may end up being only one instantiation of a process and the relevant statistics will be meaningless. Whether or not *that* is a problem may well depend on how much the reliabilist wants to rely on statistics in understanding reliability, a subject to which we shall return shortly.

The analogue of this approach on the more traditional foundationalist view would be to take again the relevant confirmation relations that exist given the evidence base *as a whole*. One such approach is relatively simple if not unproblematic. One could argue that *P* is justified for *S* if *S* is directly aware of the fact that the conjunction of propositions *S* is noninferentially justified in believing confirms *P.* Suppose that *S* is noninferentially justified in believing *E* and is aware of the legitimacy of primary inferences that allow moving from *E* to *P1* to *P2* to *P3.* Each inferential connection is legitimate but the possibility of error between each link adds up, and intuitively we want to deny that *S* has justification for believing *P3.* In such a case it is tempting to think that *E* will not by itself confirm *P3.* Confirmation, you recall, is not transitive. So the question of whether *S* has justification for believing *P3* is a question about whether the foundations stand in the confirmation relation to *P3*, something that may not be true even if there is a sequence of legitimate inferences moving from *E* to *P3.*

Whether or not one could invoke this solution to the problem of diminishing probability over sequences of inferences depends on whether or not it is plausible to claim that in the case of inferentially justified belief there always is a *direct* confirmation relation holding between the foundations and every other proposition that is ultimately justified by reference to those foundations. It may be that there is no way even in principle of moving directly from the entire foundation to every other proposition one is justified in believing. Inference rules might be like rules governing the movement of chess pieces. One can get one's knight to any square on the board, but to get it from here to there might *require* three separate moves. To get from a foundation to *P3* might similarly *require* three separate inferential moves. If this does turn out to be a major difficulty, the other alternative is to follow more closely the suggestion made in connection with reliabilism. In addition to being aware of the legitimacy of all intermediate inferences, we could require that a person be aware of the legitimacy of the *complex* inference *constituted* by the individual inferences. To be justified in believing *P3* by inferring it from *P2*, which one inferred from *P1*, which one inferred from *E*, one would need to be justified in believing the legitimacy of the inference *P3* from *P2* from *P1* from *E*, an inference that might not be legitimate even if the constitutive inferences are individually unproblematic.

Reliability

As I indicated earlier, one of the enormously attractive features of reliabilism is its *apparent* ability to secure a connection between having epistemically justified beliefs and having true beliefs. It is probably impossible for any inferential internalist who recognizes the legitimacy of nondeductive inference to secure this connection. Indeed, the present controversy between inferential internalism and inferential externalism was foreshadowed over forty years ago by another controversy about the way to understand the concept of probability relevant to epistemology, a controversy to which we return in chapter 7 when we explore the implications of internalism for skepticism.

Whether or not a reliabilist actually succeeds in connecting justification to truth depends crucially on how reliability is defined. On the crudest versions of reliabilism, the connection seems straightforward. Unconditionally reliable processes are simply those that produce mostly true beliefs. Conditionally reliable processes produce mostly true beliefs when the input beliefs are true. This straightforward statistical understanding of reliability is absolutely fatal for the plausibility of the view, however. In the preceding section we noted that belief-forming processes can be described in very general or very specific terms. There is nothing in principle to prevent us from describing a process in such specific terms that it is *in fact* employed only once. Suppose, for example, that it is a basic feature of my constitution that when I see lightning strike a bird I immediately believe that someone in my family will die. This is not a derived rule. This is, let us imagine, a belief-independent process that for some bizarre evolutionary reason operates in me. Let us suppose further that on the single occasion (past, present, and future) on which I see a bird struck by lightning, I believe my aging grandmother will die and she does. Statistically, the process has a 100 percent success rate, is 100 percent reliable. No one else (past, present, and future) uses a process like this and so we cannot turn to other people for relevant statistics. Clearly, we would not view a belief produced by such a process as justified, and so we have an objection to this most primitive form of reliabilism.[7]

Obviously, this kind of objection to reliabilism is not going to slow the reliabilist down much. It shows only what one should have realized immediately, namely, that the relevant sense of probability cannot be defined in terms of statistical regularities. The problem is directly analogous to that faced by regularity theories of law. The most difficult problem facing a regularity analysis of universal and statistical law is the problem of distinguishing lawful from mere accidental regularities. It is true that all of the people in this room are under six feet tall now, but it is not a law of nature. It is true that most of the people who read this book are philosophers, but it is not a probabilistic law. Although fraught with difficulties, a seductive approach to understanding the difference between law and accident involves turning to counterfactual conditionals. It is a law that all *F*s are *G*s when it is not only true that there are no *F*s that are *not-G* but it is also true that if anything *were* an *F* then it *would* also be *G*. It is a probabilistic law that most *F*s are *G*s only if it is true that *were* a great many things *F*, most of them *would* be *G*. The concept of reliability that the reliabilist wants is that of *lawful* probability. That is why in my earlier remarks I included the reliabilist among those externalists who seek to reduce fundamental epistemic concepts to nomological concepts.

The use of subjunctive conditionals to define a distinction between genuine reliability and accidental correlations between belief and truth is by no means straightforward, even if we have a philosophically satisfying account of subjunctive conditionals. Suppose on a space voyage to a distant planet I suffer oxygen deprivation and become very silly as a result. Immediately upon my arrival on the planet I decide *out of the blue* to infer that when an object appears round it is really square and when an object appears square it is really round. Suppose further that on this planet (call it Weird World) it just so happens that round objects and only round objects appear square and square objects and only square objects appear round. Relative to my new environment my belief-forming process concerning shapes will be nicely reliable. I not only reach true conclusions about how things are but I *would* continue to reach true conclusions *were* I to continue forming beliefs this way. But there seems to be at least some sense in which it is a kind of fantastic coincidence (accident) that my new belief-forming process is reliable. Notice that if the inhabitants of Weird World had evolved to make the same sort of judgments about shapes in the same way, we might be much more inclined to view the resulting beliefs as nonaccidentally reliable in a manner relevant to their being justified. This might suggest to some that in order for a reliable process to produce a justified belief, the very explanation for the existence of that process must involve reference to the fact that it has been successful in arriving at truth.[8]

Goldman himself seems to prefer the nebulous and metaphysically problematic concept of *propensity* as the key to understanding reliability, admitting that there is a connection between the propensity of an *F* to be *G* under conditions *C*, and the truth of the counterfactual, if a lot of things were *F* in conditions *C*, most of them would be *G*. He tells us, however, relatively little about what kind of property a propensity is. Whether one uses propensities or counterfactuals to define reliability, one must admit that one is sailing in philosophically troubled waters.

The counterfactual conditional has been a notorious thorn in the side of philosophers. There appears to be an intimate connection between the concepts of a law of nature, causation, and the relation expressed by subjunctive conditionals, but it is terribly difficult to find an analysis of one of these concepts that does not presuppose an understanding of the other two. Possible worlds analyses seem almost a joke given the need to distinguish the relevant species of possibility, distinctions that cannot be drawn without implicitly relying on an understanding of such concepts as logical and lawful possibility. Nevertheless, it would be foolish to make too much of this as a criticism of reliabilism. The nomological modalities the reliabilist needs run throughout ordinary discourse. Brief reflection will tell you that an enormous number of our familiar concepts involve dispositions, and it is hard to see how dispositions can be explicated without relying on counterfactuals or propensities. If we use the reliabilist's reliance on these concepts as ammunition to attack the view, we run the risk of leaving ourselves pitifully little to use in the construction of our own philosophical analyses.

The important thing to realize, however, is that once one turns to propensities or counterfactuals to define reliability, *one loses the connection between having justified beliefs and having mostly true beliefs.* A process can have a propensity to turn out mostly true beliefs even if in fact it almost always results in false beliefs. It will not be necessarily true given the more sophisticated versions of reliabilism that most of *my* justified beliefs are true now, that most of *my* justified beliefs past, present, and future are true, that most justified beliefs (mine and others) are true now, or that most justified beliefs (mine and others), past, present, and future, are true. And once one realizes this, one reason a philosopher might have for embracing the view disappears. We return to this matter in chapters 6 and 7 when we look more carefully at the internalist/externalist controversy in light of the skeptical challenge.

It should be added that even if the reliance on more and more specifically described belief-forming processes is not vulnerable to counterexample by virtue of the paucity of instantiations of such processes, it does invite serious epistemological problems when it comes to second-level justification, that is, justification that one has a justified belief. Even if we rely on counterfactuals or propensities and deny that these are to be defined in terms of actual statistics concerning true input/true output ratios, such statistics might be the only *evidence* one could have for supposing a belief-forming process to be reliable. And a belief-forming process with limited instantiations will not provide a statistically reliable base on which to reach conclusions. We shall return to this potential difficulty later when we discuss externalist responses to skepticism.

Before considering some of the more fundamental objections to the reliabilism just discussed and examining some of the revisions Goldman made in later work in light of these objections, let us pause to summarize the senses in which reliabilism is an externalist metaepistemology. Certainly, Goldman is a strong access externalist. He explicitly denies that having a justified belief requires that one be justified in thinking that one has a justified belief. He will also be a weak access externalist in all but very narrow conceptions of access. I cannot think of any reason why he would deny that whenever one is justified in believing *P* it is logically possible that one have justification for believing that one is justified in believing *P*. Indeed, it seems to me that he should always allow the logical possibility of having noninferential justification for thinking that one has a justified belief. After all, the notion of a belief-forming process places absolutely no restrictions on how in principle the human mind could be programmed. We could even imagine that evolution found a need for humans to have reliable second-level beliefs and so "programmed" us to believe that a process is reliable only when it is. If we were so programmed, then whenever we were to have a justified belief, we would be justified, perhaps even noninferentially justified, in believing that we had a justified belief. If a weak access requirement, that is, the possibility of access requirement, is understood in terms of what is lawfully possible, or what is lawfully possible given the present state of the subject, there is no reason to suppose that Goldman would have any interest in embracing weak access internalism. Again, however, I emphasize that this issue does not seem to lie at the *heart* of the internalism/externalism debate, for even if he sought to incorporate such weak access requirements into his metaepistemological view, it would surely stay externalist if he continued to understand the relevant access in terms of his nomologically defined reliability.

Concerning inferential justification, the reliabilist is an *inferential* externalist. Belief-dependent processes can generate justified beliefs whether or not the believer has any reason to believe that the input beliefs make probable the output beliefs. Finally, I reiterate the claim I made earlier when trying to characterize what is distinctive about traditional foundationalism. The reliabilist is trying to analyze or explicate epistemological concepts by relying on nonepistemic, "natural" concepts, specifically nomological concepts.

Fundamental Objections to Reliabilism

The most common objections to reliabilism (and externalism in general) focus on thought experiments calling into question both the necessity and the sufficiency of reliability for justification. The problem that most worried Goldman focuses once again on the familiar skeptical scenarios. Ironically, this time the old skeptical scenarios are used in an attempt to convince the reliabilist that there are justified beliefs that are not reliably produced. Goldman specifically considers Descartes's demon hypothesis in which a very powerful demon puts people in a world containing no physical objects, but produces in them the very sensations that we have, sensations that lead them naturally enough to the false conclusion that various physical objects exist. Let us suppose that the ''software'' in the minds of demon-world dwellers is precisely the same as the ''software'' in our minds, and that the input and the output beliefs in both worlds are exactly the same, the input being something like the data received through the five senses, the output being commonsense beliefs about the physical surroundings. By hypothesis, the inhabitants of a demon world will have unreliably produced beliefs; the inhabitants of our physical world (if it is as we think it is) will have reliably produced beliefs. But do we really want to characterize the victims of the demon as unjustified or irrational in believing what they do? There is an enormously powerful intuition that whatever we say by way of evaluating our own epistemic rationality, we should say about the victims of demonic machinations.

One must be very careful in evaluating one's intuitions underlying the consideration of this hypothetical situation. Most philosophers who take the thought experiment to be decisive make the mistake of appealing to the alleged normativity of justification. We could hardly *blame* the victims of the demon for believing what they do. We would have done exactly the same thing were we in their shoes. But if we can't blame them, then how can we *criticize* their beliefs by calling them unjustified or irrational? In chapter 1 we discussed at some length the claim that justification is a normative concept. I argued that while we can in a sense criticize a belief as unjustified or irrational, the criticism need not involve in any sense the assigning of blame. Indeed, I suggested that in the broad sense of criticism in which it is almost uncontroversial that evaluations of justification and rationality can involve criticism, just about any property of a belief *can* be praised or criticized provided that the person doing the evaluating holds the appropriate attitudes. I tend to criticize people who hold epistemically irrational beliefs but only because I value epistemic rationality. From the fact that the inhabitants of a demon world cannot be blamed or faulted for believing what they do, it does not follow that we cannot view their beliefs as importantly defective. They are defective in that they are false, to be sure. But they are also defective, the reliabilist could argue, in that they are unreliably produced.

When faced with the demon objection, Goldman originally hedged his bets,[9] but in *Epistemology and Cognition* he dramatically modified his reliabilism to accommodate the intuitions of his critics. The modification concerned his understanding of reliability. The most straightforward way of understanding reliability, as we saw earlier, is either in terms of statistical facts about the actual world, propensities that exist in the actual world, or counterfactuals about what would occur in the actual world were a belief-producing process used long enough. One would naturally suppose that reliability in a world is to be understood in terms of what goes on *in that world*. But in his book Goldman suggests that we understand the reliability that determines the justification of someone's belief always by reference to reliability in *normal worlds*, where normal worlds are defined by reference to certain sorts of fundamental beliefs people have about *this* world. It is never all that clear which beliefs are to be included among the fundamental beliefs. They presumably include such presuppositions of our ordinary thought as that there is a past and that our memory has something to do with it, that there is an external world and that perception is causally affected by what happens in that world, and so on. We are *not* supposed to build so much into the characterization of a normal world that it becomes analytic that any process we take to be reliable is reliable, but it seems to me, nevertheless, that it may end up being analytic that certain fundamental processes we take to be reliable are reliable. The idea behind the move to normal worlds reliabilism is that we can finally accommodate that strong intuition leading us to conclude that the victims of demonic deception are justified in holding their beliefs about the physical world. While their belief-producing processes are unreliable in that possible world, they are not necessarily unreliable in *normal worlds*.

Notice that the move to normal worlds removes reliabilism even further from the original goal of tying justification to truth. We might live in a demon world and still have justified beliefs by virtue of the fact that the processes that produce them are reliable relative to normal worlds, that is, relative to our (false) presuppositions about the fundamental nature of this world. These justified beliefs will all be false and will have been produced by processes that are terribly ineffective when it comes to getting at the truth. Unless one has some independent reason for believing that this *is* a normal world, why would the concept of justification even be important to a truth seeker when it is now so obvious that having justified beliefs need not even make *probable* having true beliefs?

Perhaps for this reason and perhaps for reasons having to do with the difficulty of distinguishing fundamental presuppositions that go into defining normal worlds from the beliefs about reliability which we want open to falsification, Goldman abandons this way of trying to assuage his demon-world critics shortly after publishing the book. He still cannot help feeling the strength of the demon world objection, but he resolves his own conflicting intuitions by recognizing a fundamental ambiguity in the concept of justification. In an article by the same name he distinguishes strong and weak justification.[10] Strong justification gets a reliabilist analysis very close to the original view presented in the early influential article. Reliability is world relativized. The reliability that determines whether or not a belief is justified is the reliability of the process that produces the belief in the world in which the belief is produced. But there is another way in which we assess the justificatory status of a belief, another way in which we can think of epistemic blame and praise. Goldman calls this weak justification and understands it essentially in terms of whether or not the belief was formed in accordance with the believer's standards:

> *S*'s belief is weakly justified at the primary level if (1) the cognitive process that produces the belief is unreliable, but (2) *S* does not believe that the producing process is unreliable, and (3) *S* neither possesses, nor has available to him/her, a reliable way of telling that the process is unreliable. . . . (4) there is no process or method *S* believes to be reliable which, if used, would lead *S* to believe that the process is unreliable. (p. 59)

The concept of weak justification is in many ways closer to that of a coherence theory, and we explore this sort of view in our more general discussion of coherence theories. For now I concern myself only with the strong concept of justification that goes back to the core idea of reliabilism.

In suggesting that we must be careful in appealing to the normativity of epistemic judgments by way of attacking reliabilism, I did not mean to suggest that I disagree with those who argue that the epistemic status of the demon victim's beliefs is the same as the beliefs we hold. Indeed, it seems *to me* virtually self-evident that there is a sense of rationality such that if beliefs in our world are rational, beliefs in the demon world are equally rational. It seems evident to me that everything relevant to assessing the rationality of the respective beliefs is present in exactly the same way in both worlds. I just think it is a mistake to go on to try to explain this intuition by reference to what we would consider praiseworthy or blameworthy. Similarly, I agree entirely with the arguments BonJour raises against reliabilism (and other versions of externalism) in chapter 3 of BonJour (1985). It seems obvious to me that a clairvoyant who has no reason to believe that her premonitions are usually accurate does not have a justified belief about the future even if she is truly clairvoyant, even if she does have 100 percent reliable clairvoyant belief-producing mechanisms. You might think this is a dangerous admission for me to make, since BonJour is a strong access internalist and that is a view I have rejected. But none of the examples he discusses are cases in which I would recognize the individual as having a noninferentially justified belief. The clairvoyant's beliefs about the future are at best inferentially justified, and since I am an inferential internalist I do hold that one must have justification for thinking that there is the relevant connection between the data from which one infers a conclusion and the truth of the conclusion inferred.

I am realistic enough to realize, however, that none of the above arguments are going to convince any serious philosopher who has carefully considered and then embraced reliabilism to abandon the view. The lines are pretty much drawn in the sand by now. The respective camps know what they have to say in order to defend their views and they are willing to say it. Concessions are made here and there. Modified reliabilism acknowledges that clairvoyants who have reason to believe that they are not clairvoyant may not have justified beliefs because of the availability of alternative legitimate reasoning which would have led them to distrust their predictions.[11] And it is in fact rather difficult to imagine a clairvoyant who would not have at least inductive evidence against the existence of such powers. Difficult, but not impossible, and the counterexample will be presented again with the hypothetical situations described so as to rule out the availability of evidence debunking claims of clairvoyant power.

Reliabilists know what they must say in order to be consistent reliabilists. It is no accident that the view gained dramatic popularity in the age of the computer. The temptation to anthropomorphize one's computer is almost irresistible, and the converse tendency to model the human brain on the workings of a computer is equally understandable. What could be more natural than to think of the brain as just a very complex computer programmed to process an enormous array of data? A computer can run well or it can run badly. It is doing its job when it performs its "calculations" correctly, when it gives the "right" response to queries and problems. The epistemic computer that is our brain is doing well when it represents correctly the world from which it receives input. It is behaving improperly (and is to be criticized accordingly) when it habitually calculates incorrectly the character of the surrounding world. Effective computers do not have to know that they are effective. When a computer makes an "inference" from a set of "input" truths to a "conclusion," and gets things right, it is doing its job whether or not it can respond correctly to inquiries concerning the connection between the input and the output. Similarly, the human mind is doing its job when it moves from truth to truth whether or not it has the slightest conception of what makes the move truth-preserving.

Indeed, it seems to me that even die-hard internalists should admit that the distinction emphasized by the reliabilist between reliable and unreliable belief-forming processes is a perfectly intelligible and even useful distinction to make. I suspect that it is even a distinction that occasionally finds expression in ordinary discourse using the terms of epistemic evaluation. It may be that it is particularly natural to focus on questions of reliability in contexts where one is making second-and third-person judgments of epistemic justification or rationality. Consider the now famous example of the chicken sexer. It is alleged that there are a few unusual people who can make a very good living by virtue of the fact that only they can tell whether a baby chick is male or female (something that is for various reasons useful to know as soon as possible). What is philosophically interesting about these people is that they are not supposed to have the faintest idea *how* they tell the male from the female chicks. They look at the chick and they just find themselves believing either that it is male or that it is female, and their beliefs are almost always correct. They have, that is, some kind of reliable belief-forming process at work but nobody knows exactly what the process is or how it works.

Now, to be honest, I seriously doubt that there are people who satisfy this description. If I were one of a select number of people who knew how to tell male from female chicks and the skill was worth a lot of money, I suppose I too would pretend I did not know how I was doing it. But there *could* be such people and the philosophically relevant question is whether we would allow that our hypothetical chicken sexer knew or had a justified belief about the sex of the chick. And so that we do not confuse matters, we should ask the question of the chicken sexer's beliefs before she accumulated enough successful predictions to form the basis of an inductive justification for her belief. It would not surprise me in the least if most would conclude that the chicken sexer knew all along which chicks were male, and which female. And I suspect that what they mean when they grant the chicken sexer knowledge is best explained along reliabilist lines (or one of the other familiar externalists views). But if we imagine the chicken sexer, on the occasion of her first belief while holding before her a baby chick, waxing philosophical and wondering whether this funny belief that popped into her head was justified, it is hard to imagine that the mere existence of the relevant nomological connections is the object of her concern. When she wants to have justification for this belief, it is difficult to imagine, in other words, that what she wants is merely to be programmed so as to respond correctly. To get justification now for the belief is not to be in a situation in which one stands merely in certain nomological relations to stimuli.

I am not arguing that concern with the reliability of belief-forming processes is restricted to second- and third-person epistemic judgments. I have admitted that the reliabilist has drawn a distinction that is legitimate and useful. If I had to choose between being a person with reliably produced beliefs or being a person with unreliably produced beliefs, I suppose I would choose the former. It would be better for all sorts of reasons. But in my role as philosopher I would not be satisfied with having reliably produced beliefs, or even with having reliably produced beliefs about which of my beliefs are reliably produced. I am going to argue, when exploring the implications for skepticism of a view like Goldman's, that there are perfectly natural and legitimate questions that a *philosopher* is interested in that cannot be answered employing epistemic concepts analyzed in terms of reliability. It is going to be difficult (probably impossible) to convince the reliabilist of this. As I indicated, the lines are drawn in the sand. If one is going to talk people out of inferential externalism, I think it will have to be through the use of a much more subtle sort of philosophical persuasion. I shall try to convince the reader that the very ease with which the reliabilist can respond to epistemological questions *at all levels* will, ironically, convince many that there is a kind of epistemological question we want to ask that the reliabilist cannot formulate given his analysis of epistemic concepts.

Nozick's Analysis of Epistemic Concepts

I have focused so much on Goldman's externalism because I think it is the most plausible, sophisticated, and influential of the paradigmatic externalist views. Nozick's metaepistemology has also had considerable influence, particularly in connection with his response to traditional skepticism, and so I briefly attempt to indicate the way in which I think the general conclusions I reached concerning externalism apply to Nozick's view as well.

Nozick's epistemology is best known for his account of knowledge rather than his views about justified belief. Indeed, he seems to simply accept a version of reliabilism as adequate to understanding the concept of epistemically justified belief. He says that a belief that *P* is justified if it was arrived at via some method *M* such that that method is "likely to produce mostly true beliefs" (Nozick 1981, 264). I am not sure what the expression "likely" is doing in this externalist analysis of justification. Either it is redundant (its meaning exhausted by the reference to producing mostly true beliefs) or it is an epistemic concept surreptitiously introduced into an analysis that is supposed to be purged of epistemic terms. I assume, however, that, as with Goldman, the relevant processes conferring justification are those that usually produce true beliefs (if they are unconditionally reliable), or would produce mostly true beliefs if they were used indefinitely, or have a propensity to produce mostly true beliefs.

Nozick modifies his reliabilism for reasons similar to those motivating Goldman to introduce the relevance of alternative methods. When a number of reliable methods are available to an individual vis-à-vis reaching a conclusion about *P*, the belief will be justified only if the most reliable method is used. On the face of it, this suggestion will run into difficulties. Suppose that there are five methods, *M1* through *M5*, that I could use in deciding whether or not *P* is true. *M1* is the most reliable of the five and leads to the conclusion that *P*. Each of *M2* through *M5* leads to the conclusion that *not-P*. Obviously, a number of different, less reliable processes leading to the opposite conclusion can more than match a single process leading to *P*. The same suggestion I made to Goldman would be equally applicable here. Rather than think in terms of the impact of individual methods used, why not employ hypotheticals about what conclusion would be reached were all methods used simultaneously?

I will return to Nozick's views about justified belief in a moment. As I indicated, however, the most discussed aspect of Nozick's metaepistemology is his "tracking" analysis of knowledge: *S* knows that *P* when *S*'s belief that *P* tracks the fact that *P*. The analysis of tracking gets progressively more sophisticated in order to deal with obvious objections. The crude idea is that *S*'s belief that *P* tracks the fact that *P* when (1) if *not-P* were true, then *S* would not believe *P*, and (2) if *P* were true, *S* would believe *P.* Nozick considers cases in which these conditions would not be satisfied but in which we would still want to allow knowledge. He describes a mother who upon seeing her son concludes naturally enough that he is alive. Had her son died, however, her friends would have conspired to keep the information from her. She would still, therefore, have believed that he is alive even if he had died. To accommodate the intuition that the mother had knowledge *based on perception*, Nozick sees that he must make reference to the method by which a conclusion is reached:

> *S* knows that *P* when:
> 1. *P* is true
> 2. *S* believes via some method *M* that *P* is true
> 3. if *not-P* were true and *S* were to use *M* to arrive at a belief about *P,* then *S* wouldn't believe, via *M,* that *P.*
> 4. If *P* were true and *S* were to use *M* to arrive at a belief about *P,* then *S* would believe, via *M,* that *P.*

Although the intuitions are not as strong, Nozick feels he needs to add one more qualification in his account to deal with cases in which a belief is causally overdetermined by the use of more than one method. If more than one method is used to support a belief that *P* and one of the methods fails to satisfy tracking requirements *and* that method *outweighs* the method that does satisfy tracking requirements, Nozick wants to say that the person does not know. He gives the example of the father who believes in the innocence of his son on perfectly good evidence even though he also would have believed in the innocence of his son, on the basis of blind faith. If the faith dominated the reliance on evidence as the causal factor sustaining belief, then the father did not know that his son was innocent. To get this result, Nozick adds a fifth condition to the analysis of knowledge:

> 5. There is no other method which outweighs *M* such that conditions 3 and 4 are not satisfied by that method.

M is outweighed by other methods if ''when *M* would have the person believe *P,* the person believes *not-P* if the other methods would lead to the belief that *not-P,* or when *M* would have the person believe *not-P,* the person believes *P* if the other methods would lead to the belief that *P*'' (p. 182).

Although Nozick recognizes the relevance of alternative methods that *causally sustain* belief to the question of whether the believer has knowledge, one might argue that condition 5 needs to be modified still further. Recall that Goldman worried about the man who has a reliably produced belief about the past but has ample evidence available that should have led him to distrust the reliability of his memory. As I understand Goldman's description of the situation, the man's belief satisfies Nozick's tracking requirements for knowledge. There are no other methods of inference causally sustaining the man's belief that fail to satisfy the tracking requirements for knowledge and that outweigh the method used. But surely the availability of methods that could have been used, should have been used, and would have led to contradictory conclusions, is relevant to the question of whether the belief in question constitutes knowledge. Condition 5 could be modified to include reference to all available methods (where the use of more than one method could be construed simply as the use of a complex method).

Although it is not something emphasized by Nozick, it seems to me that like reliabilism, Nozick's analysis of knowledge is foundational in structure, or at least his analysis of knowledge can accommodate a distinction between inferential and noninferential knowledge. Methods that track facts can presumably be either belief-dependent or belief-independent. Beliefs based on memory or perception, for example, might be construed as beliefs that do not involve inference from other propositions believed. Other methods, like deductive and nondeductive inference, do involve (at least implicitly) formulating premises for use in arguments. Furthermore, I assume that there is at least a sense in which all of the belief-dependent methods that are successful in tracking facts depend ultimately on belief-independent methods that are successful in tracking facts. Otherwise it is difficult to see where one would get the premises for use in the effective belief-dependent methods. Like the reliabilist, and virtually all other self-proclaimed externalists, Nozick will presumably deny the second clause in the principle of inferential justification, or at least will deny its analogue for inferential knowledge. One can know that P by using a belief-dependent method that is successful in tracking the fact that P even if one has not the slightest reason to suppose that the belief does track the relevant fact nor the slightest reason to suppose that the method was effective in securing the tracking relation.

The tracking analysis of knowledge, like a reliability theory of justification, attempts to reduce epistemic terms to nomological terms. Nozick employs the possible worlds metaphor in trying to say a few helpful words about the subjunctive conditionals he uses: it is true that if X were the case then Y would be the case when in the close possible worlds in which X is the case Y is the case. But to his credit he realizes that it is silly to try to define subjunctive conditionals using the concept of possible worlds. As he himself puts the crucial point, the only measure of closeness or similarity that the *not-P* worlds must have to the actual P world in order to be relevant in assessing the truth of subjunctive conditionals with *not-P* as an antecedent, is that the *not-P* worlds be just like the actual world except with respect to the changes that *would* have occurred in the actual world had *not-P* been the case. If one recognizes that subjunctive conditionals cannot be analyzed using the metaphor of possible worlds, the most plausible alternative involves turning to the concept of *law*. I have suggested elsewhere (Fumerton 1976) that we might begin trying to understand *contingent* subjunctive conditionals of the sort Nozick needs for his analysis this way:

> If P were the case then Q would be the case iff there is some law L such that (P and L) entails Q where neither P nor L alone entails Q; or there obtains some state of affairs X and some law L such that (P and X and L) entails Q, where (X and L) alone does not entail Q and where X does not contain P as a part.

Y is a part of X if X contains Y, or X contains some state of affairs that contains Y, or X contains some state of affairs that contains some state of affairs that contains Y, or . . . . X contains Y if X is identical with the state of affairs that is Y being related to one or more states of affairs via truth-functional connectives.

The difficulty with the above "analysis" is that in asserting subjunctive conditionals we often want our audience to imagine changes in addition to the specific change mentioned in the antecedent in evaluating the conditional. At the same time it is not clear that there are any hard and fast "rules" governing what changes we are to take into account. If I ask you what would have happened after Iraq had invaded Kuwait were Cyrus Vance secretary of state, are you to imagine Vance occupying that position while Bush was still president? Or in imagining Vance secretary of state are you to imagine that Dukakis won the contest with Bush? It is not implausible to suppose that the most common rule we follow in ordinary discourse involves imagining all those changes made *likely* by Vance's being secretary of state. But if "likely" here is an epistemic notion (as it almost certainly is), then the use of the kind of subjunctive conditionals invoked by Nozick to analyze epistemic concepts might very well involve disguised circularity.

It would be a mistake to worry too much about Nozick's use of the subjunctive conditional, given the notorious philosophical difficulties one encounters in attempting to analyze them. My primary purpose is to make clear that contingent subjunctive conditionals appear to invoke implicitly the nomological concept of lawful necessity, and consequently Nozick's account of knowledge fits my earlier characterization of the externalist as one seeking to reduce epistemic concepts to nomological concepts.

There all kinds of objections one could raise against Nozick's tracking conditions for knowledge. The analysis encounters a number of difficulties, for example, in trying to accommodate knowledge of necessary truths. Suppose I know via some method M a necessary truth N. Is it even possible to evaluate counterfactuals with necessarily false antecedents? Is it possible to reach a conclusion about what I would have believed had N been false? Because he is skeptical about the intelligibility of such a question, Nozick suggests a simplified analysis of knowledge for necessary truth. He suggests that you know a necessary truth N via some method M if in all of the close possible worlds in which you employ M you believe that N. Given this view, it would unfortunately seem to follow that if I resolve to believe every mathematical proposition I entertain (this is my ''method'' of arriving at a belief), I will know all of the necessarily true mathematical propositions (no matter how complex) I entertain, provided that I happen to be so constituted that I would entertain these propositions in all ''close'' possible worlds! If he thought more about it, I suspect that Nozick would retreat in the face of this objection to a reliabilist account of knowledge for necessary truths. But if the externalist needs reliabilism to analyze knowledge of necessary truth, that is surely a reason ceteris paribus to prefer reliabilism as a general account of knowledge and justification.

In one of his most daring and dialectically ingenious philosophical moves, Nozick takes one of the most counterintuitive consequences of his view and embraces it as a virtue. It follows from the tracking analysis of knowledge that I can know E and know that E entails P, believe P as a result of believing E and that E entails P, and yet not know that P. Knowledge, for Nozick, is not closed under known implication. Rather than take this to be a reductio of his account of knowledge, Nozick claims that he can accommodate the skeptic's intuition that we cannot rule out skeptical hypotheses while retaining the commonsense view that we know commonplace truths about the world around us. I can know through perception that there is a table before me now by virtue of the fact that my belief that the table exists tracks its existence.

If the table were not there, then I would not believe in its existence on the basis of visual experience. I cannot know that I am not being deceived by an evil demon with respect to the existence of a table, for if I *were* being deceived by an evil demon with respect to the existence of the table, I would continue to believe that I am not being deceived. Intuitively, all of the skeptical scenarios are drawn in such a way that were they true or false, I would go on believing what I do about them since their truth or falsehood would not *reveal* itself to me in any way that I could take into account.

Looked at as a way to resolve the tensions between giving both common sense and the skeptic their dues, Nozick's tracking account of knowledge is certainly ingenious. It should perhaps be emphasized that the failure of knowledge to be closed under known implication might also be a feature of the much older causal theories of knowledge. On a very primitive causal theory, if the fact that P caused me to believe P, then I know that P. If P is a fact, it follows that its not being the case that Q caused P to be false is a fact. I might even believe as a result of deductive inference that it is false that Q caused P to be false. But it is not clear that this more complex fact was a link in a causal chain leading to my belief in the proposition asserting the absence of that causal relation.

If Nozick's denial that knowledge is closed under known implication has its advantages, it also has its disadvantages. The primary disadvantage is that the conclusion is absurd on the face of it. Like the reliabilists who know what their reliabilism commits them to, Nozick is unlikely to be goaded into giving up his view in the face of any alleged counterexample. The most we can do is make him as uncomfortable as possible. Nozick wants it to be the case, for example, that one can "almost" always know that there exists an F by inferring it from one's knowledge that a is F (p. 236). In a discussion relegated to a footnote (p. 693, n. 68), however, he acknowledges that in exceptional cases even this sort of proof might fail to yield knowledge. Let us describe such a case.

Suppose that I know, in Nozick's sense of the term, that Jones murdered Smith. I know it as a result of reading it in a newspaper that accurately recorded the event. Suppose further that Jones was part of an extremely sophisticated conspiracy and that the plan to assassinate Smith involved a complicated series of backup plans. If Jones had failed, there was another assassin down the hall who would have murdered Smith. And if this assassin had failed, there was a third who have murdered Smith, and so on down a line of twenty ready and able assassins. Had one of these other assassins murdered Smith, the newspaper would have accurately reported the fact that Jones was not the killer. However, in the *extremely unlikely* event that all of the assassins failed, someone trusted by the local papers would call in the false report that Jones had assassinated Smith so as to spread the sort of confusion that might help them make their getaway. As I understand Nozick's analysis of knowledge, I would know that Jones murdered Smith, if it were true, because my belief that Jones murdered Smith would track the fact that he did. In all of the ''close possible worlds'' in which Smith was not murdered by Jones (that is, the worlds in which one of the other assassins murdered Smith), I would not believe on the basis of a newspaper story that Jones murdered Smith. I would not know, however, that someone murdered Smith, that Smith was murdered, because if *that* proposition were false I would still believe that it was true by reading the false story that would have appeared in the paper.

Does anyone really want to allow that I could *know* that Jones murdered Smith without knowing that Smith was murdered? I am sure that Nozick does—this is probably the kind of exceptional case he had in mind. But the hypothesis that one could know that Smith was murdered by Jones without knowing that Smith was murdered is so absurd that it seems to me we really must abandon an analysis of knowledge that leads to this result. Again, I realize that this will not convince everyone, and when we look at externalist attempts to deal with skepticism I will return to Nozick's analysis and look again at the position he really leaves us in when it comes to resolving skeptical issues. If there is another kind of argument against the view, it will again involve (as it will with reliabilism) the way in which the view becomes increasingly more odd as we move up levels.

Although his reliance on the nonclosure of knowledge under known implication can produce strongly counterintuitive results, there is at least one place in which he would have been welladvised to insist on it. Nozick points out that on his account of knowledge, if one knows that P and one knows that Q one will know that P and Q (p. 236). While it seems right to me that this should be the case for knowledge (as opposed to having a justified belief where the principle is obviously false), it presents difficulties for Nozick's account of knowledge. It is a well-known policy of airlines to overbook flights. If they do not *know* that at least some of the passengers booked will not show up, they certainly have good reason to believe it. Suppose now that there is a naive and not very bright new employee of the airline who after looking at the reservations concludes that each passenger $P1$ through $P200$ has arrived at the airport ready to board the plane. The employee is panicked at this thought, because he realizes that the plane can carry only 195 passengers. Let us assume further that by a remarkable coincidence his beliefs with respect to each booked passenger that he or she has arrived are true and that, furthermore, each belief tracks the fact that makes it true. He just happened to have beliefs about two hundred people none of whom would fail to arrive without prior notification in any but the remotest of possible worlds. I take it that on Nozick's account of knowledge, the naive employee would know of each booked passenger that he or she had arrived and would consequently know that they had all arrived. But surely this is wrong. The naive airline employee had an irrational true belief that all of the passengers showed up and *for that reason* did not know that they would all show up. There was available to the employee all kinds of very good evidence from which he should have inferred that his belief was likely to be false. Although tracking seems like a very strong requirement for knowledge, the use of subjunctive conditionals to explain it, with the reliance on ''close possible worlds'' to explain what those conditionals assert, seems to end up making the requirement too weak.

Before leaving the question of closure, it should be noted that nowhere in his discussion of skepticism does Nozick give his views about the possibility of *justifiably* believing that the skeptical hypotheses are false. Given his reliabilist conception of justification, it seems that Nozick is committed to the view that while we cannot know that the skeptical hypotheses are false, we can have very strong justification for believing that they are false—for believing, say, that we are not being deceived by an evil demon with respect to the existence of the table. After all, Nozick thinks that we do know that we veridically see the table, and he is inclined to think that knowledge presupposes the existence of a reliably formed belief (p. 267). I assume, therefore, that he thinks we have a justified belief that we veridically see a table. I further assume that if there are any belief-dependent conditionally reliable processes, deductive inference is one of them and that if I deduce that I am not being deceived by an evil demon now from the proposition that I am veridically seeing the table, the reliability of the latter belief will be transferred to the former. Indeed, it seems to me that on Nozick's account it may well be the case that I have an extraordinarily well justified belief that the skeptical hypothesis is false. The only explanation I can think of for why Nozick does not make this claim is that he is very anxious to give the skeptic his due and he implicitly realizes that it is hard to do that while insisting that the hypothesis you cannot know to be false is one that you could have extraordinarily strong justification for rejecting.

In his discussion of the way in which one can reconcile the force of the skeptic's claim that we cannot rule out skeptical scenarios with our commonsense view that we know the world around us, Nozick for the most part seems to simply assume that our commonsense beliefs based on memory, perception, induction, and so on *do* track the facts that would make those beliefs true. And it is tempting to suppose that Nozick simply begs the question against skepticism in making this supposition. Like the concept of reliably produced beliefs, the concept of beliefs that track facts might also be useful (provided that we can make tracking a suitably precise notion). But why is Nozick so confident that whereas our beliefs about skeptical hypotheses do not track the relevant facts, our beliefs about the external world *do* track the relevant facts? Externalists interpret this as a second-level question about knowledge or justified belief and urge us not to confuse the answer to this question with the issue of whether we have first-level knowledge. But even if we make the distinction, it is a perfectly legitimate question worth asking, and it is *tempting*, though ultimately mistaken, to suppose that the old skeptical problems will rearise at the next level. But just as the reliabilist will have no difficulty, in principle, supposing that we have reliably produced beliefs about which of our beliefs are reliably produced, Nozick need have no difficulty, in principle, allowing that we have beliefs that track facts about which beliefs track facts. The second-level problem of knowledge or justified belief, the question of how we know that we know or justifiably believe that we have justified beliefs, *is no more difficult for the externalist to answer than the first-level questions.* I know this does not seem right, but that is only because externalism, from a certain *perspective*, doesn't seem right. This is the issue to which I return when we examine closely the nature of the externalist's approach to traditional skeptical arguments.

Summary

I have looked at two paradigmatic externalist analyses of epistemic concepts, with the primary goal of understanding the underlying commitments of the views and understanding the character of the views that is essential to their externalism. At the same time, where appropriate, I noted internal difficulties that the views face and occasionally suggested modifications of the views that make them more plausible. I chose Goldman's reliabilism and Nozick's tracking account of knowledge because I think these are two of the more sophisticated and influential of the externalist analyses. Most externalist views owe a heavy debt to Goldman, and his is the most promising of the paradigm externalisms from which to generalize. It is a tribute to his influence that there are so many other variations on externalism, and by ignoring these I do not mean to denigrate their importance. At the level of generality I am interested in for our discussion of skepticism, I think there is little danger that the relatively minor differences between externalist accounts will be significant for the points I want to make.

I have not tried to convince you that either of the above paradigm externalist accounts is fatally flawed. In fact I do not think they are. I think both Goldman and Nozick provide an analysis of an interesting and potentially useful distinction—useful for at least some purposes in some contexts. I do not think either of them has analyzed a concept that enables us to ask the philosophical questions in epistemology that are legitimate and that have dominated the history of epistemology. In general, I do not think the externalist succeeds in analyzing a *philosophically* satisfying concept of knowledge or justified belief. I do not pretend to have shown this yet. If it can be shown, it will be in the context of examining the implications of externalism for the way in which we should approach traditional skepticism.

Notes

1. For an illuminating discussion of the historical question see Kim 1988.
2. Goldman 1979, p. 20.
3. The general problem of reconciling determinism with claims about what I could have done differently is no doubt familiar to you in the context of the well-known attempts to reconcile determinism with some sense of freedom.
4. See Goldman 1986, sec. 5.7.
5. See Goldman 1979, p. 13, n. 10.
6. This suggestion was made to me by a student of mine, Jim Sloan. I found his thoughts on this matter very helpful.
7. Chisholm (1989, 78) presents this sort of objection to reliabilism. I am not sure that Chisholm's objection will be effective against the more sophisticated versions of reliabilism I discuss later. See also Pollock's objection to process reliabilism relativized to circumstances (1987, 116–21).
8. Or one could introduce a broader notion of appropriate function as Plantinga (1993b) does to distinguish those reliable processes that are functioning appropriately from those that are not. One can then go on to restrict the processes relevant to justification to those that are functioning as they are supposed to. Sosa (1991) also has an interesting discussion of the problem of accidental

reliability. As I understand Sosa, the key to restricting the kind of reliability relevant to *knowledge* is to require that the knower have true, coherent meta-beliefs concerning the *source* of reliability. This requirement may, in effect, achieve the same goal as the attempt to more straightforwardly require that the reliability came about in some appropriate way.

9. See Goldman 1979, pp. 16–17.

10. Goldman 1988.

11. See Goldman 1988, p. 112.

Chapter Five

Coherence Theories of Truth
and Justification

In chapters 3 and 4 I examined paradigm versions of internalism and externalism that are foundationalist in structure. They either presuppose, or at least allow the possibility of, a belief's having a justification that does not require the having of other beliefs. We found that there is not just *one* internalism/externalism debate; rather, one can define the controversy in a number of different ways. I argued that within foundationalist metaepistemologies, the most illuminating way of understanding the distinction might be in terms of whether one is committed to the possibility of reducing epistemic concepts to nonepistemic concepts, specifically nomological concepts.

Historically, the coherence theory of justification is the leading alternative to foundationalism, and with the rise of the internalism/externalism controversy the coherence theorist has been forced to choose sides. In this chapter I examine coherence theories of justification and the coherence theories of truth that sometimes underlie them. With our distinctions between different versions of internalism and externalism we shall attempt to fit coherence theories of justification into the framework we have set up. As was true of the discussion of foundational externalism, my examination of coherence theories of justification will be critical, but once again one of my primary interests is to explore the implications of the theory for the way in which it will respond to skeptical arguments.

The most natural theory of truth for virtually all foundationalists, internalists and externalists alike, is a correspondence theory of truth. Acquaintance theories, causal theories, reliabilist theories, and tracking theories of noninferential knowledge or justified belief seem to implicitly suppose that there is a world out there logically independent of our representations of it, but capable of being correctly or incorrectly represented by belief. The relevant knowledge or justified belief comes about when we are suitably related to that world. There are enormous differences among foundationalists when it comes to understanding the nature of the relevant relation and, to be sure, it makes a huge difference whether it is a primitive, unanalyzable relation of acquaintance, a causal connection, or a stimulus processed by a brain programmed to churn out representations of the world when affected with such a stimulus.

In suggesting that foundationalism seems to go most naturally with a correspondence conception of truth, I am not arguing that there would be any formal inconsistency in combining a coherence theory of truth with one of the externalist conceptions of justified belief. I suppose there is nothing to stop a philosopher from defining justified belief in terms of reliably produced belief, and then understanding the truth of the proposition that a belief is reliably produced in terms of the coherence of that proposition with some set of other propositions believed. But if truth really is to be understood in terms of coherence, by far the most natural view of justification would be a coherence theory of justification. If I want epistemic reasons to believe P when the truth of P consists in its cohering in a certain way with other propositions, what better reason could I have than my awareness that this sort of coherence obtains?

Whereas I think it is true that a proponent of a coherence theory of truth will most naturally turn to a coherence theory of justification, the reverse is not as obvious. There are many proponents of a coherence theory of justification who will have nothing to do with a coherence theory of truth. But since it is an open question how much each theory will need to rely on the other, it is worth beginning our examination of a coherence theory of justification with a discussion of the coherence theory of truth.

Coherence Theories of Truth

Coherence theories of truth have never really been that popular among analytic philosophers, but like the fabled phoenix, the view keeps coming back from its ashes, though often in disguise. For example, it seems to me that the realism/antirealism debate, as it is carried on in both philosophical and pseudophilosophical circles, can some times be best understood as a debate over the plausibility of a coherence theory of truth.[1]

But what exactly is a coherence theory of truth? It might be useful to begin by reminding ourselves what it is not. One is not a coherence theorist or an antirealist of some kind just because one holds that there are no truths without minds and their representations. In chapter 3, I sketched what I take to be a paradigm correspondence conception of truth, and I specifically argued for the view that truth presupposes representation and representation presupposes minds. A correspondence theorist can make perfectly good sense of the claim that without thought there would be no truth. But thought, for the correspondence theorist, will be only a necessary condition for truth. What makes a thought true, when it is true, is the fact that the relevant relation holds between the thought and the fact that it represents. I argued for the view that this relation is a sui generis, unanalyzable relation of correspondence, but one can, of course, develop naturalistic analyses of representation, and consequently, of correspondence. Thus, if a sign always signifies what it does because of some causal relationship between the thing signified and the sign, then a correspondence theory of truth might hold that a belief that *P* is true when it has the "appropriate" causal origin. False beliefs or thoughts, though they will have some cause, presumably will fail to stand in the *appropriate* causal relationship to a fact that it could represent. It will obviously be exceedingly difficult to define representation in terms of causal origin, and proponents of the view are notorious for the sketchiness of their analyses.

The fact that minds are necessary for truth, then, does not entail antirealism or a coherence theory of truth. Nor does the fact that one is an idealist in the Berkeleyan sense of idealist. It is more than a little unfortunate that the terms "idealist" and "realist" are used in so many different ways in philosophy. Let us say that a Berkeleyan idealist believes that the only things that exist are minds and their properties. One can hold such a view and embrace a correspondence theory of truth. Some of the properties of a mind could be thoughts capable of representing. These thoughts might be true or false in virtue of their corresponding to facts. Since according to the Berkeleyan idealist the only facts that exist are facts about minds and their properties, it will follow that true thoughts will correspond to mental facts. But as long as one thinks of the truth-maker and the thought made true as distinct *and* one thinks that it is a relationship of correspondence holding between the two that constitutes truth, this idealist can be a paradigm proponent of a correspondence theory of truth. Although Berkeley does not discuss the issue in contemporary terms, I am convinced that he would have no sympathy at all for a coherence theory of truth (or justification).

So the proponent of a correspondence theory of truth and a realist *with respect to truth* can hold that the bearers of truth value are all mind-dependent thoughts and even that the facts that make those thoughts true are themselves mind-dependent. The only question relevant to whether or not one is a correspondence theorist of truth is one's analysis of the *relation* that determines truth.

Failure to be clear about this point can cause endless confusion. It is particularly important to keep things straight when the *subject matter* of an assertion is itself coherence. Obviously, a correspondence theorist will not deny that there are truths *about* coherence. It will be tempting and *in a sense* true to say that a truth asserting of two propositions *P* and *Q* that they cohere is made true by a relation of coherence. But even in this case the correspondence theorist will argue that it is a relation of correspondence between the bearer of truth value and these relations of coherence holding between propositions that makes the proposition in question true. The facts about the world that have the potential to make propositions true *include* facts about relations of coherence obtaining. We might illustrate the point with two quick examples.

It has always seemed plausible to me to suppose that literary talk about fictional characters should best be construed as assertions about what is in the text. But it is probably implausible to suppose that when the literary critic argues that Hamlet loved his mother, the assertion is intended to stand or fall on finding a sentence to that effect in the play. It is enough that Hamlet's loving his mother *coheres* well with the explicit statements and descriptions of behavior contained in the text. We might describe this theory of how to understand the truth conditions of the literary critic's utterance as a kind of coherence theory, but it should be clear now that such a theory is not a coherence theory of *truth*. A correspondence theorist can easily embrace the theory by taking the truth conditions of the literary critic's assertion to include these facts about coherence where the truth consists in correspondence of the proposition to the relevant facts.

Or consider so-called coherence theories of law. Joseph Raz (1992) discusses what he calls a coherence theory of law as if it were a species of a coherence theory of truth (a coherence theory of truth with respect to law). The rough idea behind a coherence theory of law is that the law of a certain country consists of "the most coherent set of normative principles which, had they been accepted as valid by a perfectly rational and well-informed person, would have led him, given the opportunity to do so, to promulgate all the legislation and render all the decisions which were in fact promulgated and rendered in that country" (Raz

1992, 290). Now whatever advantages or problems may attach to a coherence theory of law, it is again crucial that one *not* think of a coherence theory of law as a coherence theory of *truth* with respect to what the law is. Once again, one can embrace the aforementioned theory within the framework of the correspondence theory of truth. A proposition asserting that the law of country X is L can be thought of as corresponding to the complex facts of coherence thought by the coherence theorist of law to determine what the law is. The fact that an assertion about law might be an assertion about relations of coherence is in no way relevant to a theory of truth. Any philosopher with any theory of truth could embrace this view about the nature of law within the framework of that conception of truth.

When it comes to coherence theories of truth and justification, one can distinguish ''pure'' and ''impure'' versions of the views. The pure versions offer completely general accounts of truth and justification. The impure versions restrict their coherence theories to a subclass of truths or justified beliefs. I begin by looking at a pure coherence theory of truth. The proponent of such a view, as the label implies, thinks that truth *consists* in coherence. The two most obvious questions to ask a coherence theorist are: (1) What is it that stands in the relation of coherence? and (2) What does coherence involve?

Let us address the first question. The most obvious answer, I suppose, is that the relata of coherence for a coherence theorist are the bearers of truth value, whatever they happen to be. So one could take the relata of coherence to be sentences, statements, propositions, beliefs, or anything else that it makes sense to describe as true or false. Suppose the coherence theorist picks propositions as the primary bearers of truth value (and do not be concerned for the moment about what a proposition is). What makes the proposition P true? That it coheres with other propositions. But *which* other propositions? Is P true if it coheres with all other propositions or merely a subclass of all other propositions? Since some propositions are true and some propositions are false, it would be absurd to suggest that P is true iff P coheres with the class of *all* other propositions. The class of all other propositions includes *not-P*, and P will not cohere with *not-P*. If coherence is understood this way, nothing will be true. Is P true if it coheres with the class of all other *true* propositions? Maybe, but it does not take a master metaphysician to see that this will hardly do as an analysis of truth. Circles do not get more vicious than this.

If we are going to restrict the class of propositions with which P must cohere in order to be true, and we cannot select the relevant class by reference to the truth value of propositions, it seems obvious that we have no alternative but to use *beliefs* as the way in which to restrict the relevant propositions. *P* is true iff *P* coheres with some set of propositions that are or would be believed by the relevant person or persons. One can still think of the crucial relations of coherence as holding between propositions, but the relevant propositions with which *P* must cohere are defined in terms of their being the actual or hypothetical objects of belief. One gets different versions of a coherence theory of truth depending on how one uses belief to restrict the relevant class of propositions with which a proposition must cohere in order to be true. If the relevant subclass of propositions can vary from one individual or community to another and it is coherence with an individual's or community's belief system that determines truth, the coherence theorist will presumably *relativize* the concept of truth. Among others, one can distinguish the following coherence theories of truth:

1. The partially relativized (to time only), world, present actual belief, theory: *P* is true at *t* iff *P* coheres with all or most of what is believed at *t*.
2. The nonrelativized, world, hypothetical belief, theory: *P* is true iff *P* coheres with what all or most people would believe at the end of an indefinitely long idealized process of inquiry.
3. The fully relativized (to communities and time), community, present actual belief, theory: *P* is true at *t* in community *C* iff *P* coheres with the beliefs held by all or most of the members of *C* at *t*.
4. The partially relativized (to community, not to time), community, hypothetical belief, theory: *P* is true in community *C* iff *P* coheres with the conclusions that the members of *C* would reach at the end of an indefinitely long idealized process of inquiry.
5. The fully relativized (to individuals and time), individual, present actual belief, theory: *P* is true at *t* for *S* iff *P* coheres with the rest of what *S* believes at *t*.
6. The partially relativized (to individuals, not to time), individual, hypothetical belief, theory: *P* is true for *S* iff *P* coheres with the rest of what *S* would believe at the end of an indefinitely long idealized process of inquiry.

Those of us who occasionally teach philosophy to undergraduates have spent endless hours patiently trying to get our students to stop talking about things being true for one person but not for another. It is almost always the case that when students use this locution they really are referring just to the unproblematic fact that the same proposition can be believed by one person and not by another. But a coherence theory of truth actually gives one a philosophically respectable way of making sense of this "true for" talk. Indeed, if it is the coherence of a proposition with an individual's belief system that makes it true, then we must relativize the concept of truth and think of the proposition as true only "for that person."

Once the coherence theorist of truth settles on the appropriate relata of coherence, the next task is to specify the relevant relations of coherence. Minimally, coherence will involve logical consistency. On the conception of truth relative to an individual's belief system, *P* can be true relative to *S*'s belief system only if *P* is logically consistent with the conjunction of other propositions *S* believes. As BonJour pointed out in connection with a coherence theory of justification, logical consistency is a notoriously weak sort of coherence, and a coherence theorist of truth will no doubt want to expand the kind of coherence that yields truth to include at least probabilistic connections. *P* will be true for *S* only if *P* is logically consistent with the rest of what *S* believes, and there are interesting probabilistic connections between other propositions *S* believes and *P*. Since we will need to look much more carefully at different kinds of coherence when discussing a coherence theory of justification, and since my criticisms of a coherence theory of truth will not focus on the details of the kind of coherence invoked by the theory, I think we need not worry any more now about how precisely a proponent of the coherence theory of truth will define coherence.

A coherence theory of truth is completely unintelligible. It falls victim to a vicious conceptual regress. Philosophers have been aware for some time that proponents of a coherence theory of truth have a very awkward time applying their theory of truth to the theory itself. We have (at least) two theories of truth: a correspondence theory of truth and a coherence theory of truth.[2] The coherence theorist seems to take one of these theories to be *true* and the other false. But given most relativized versions of the theory, there seems no reason in principle to deny that a correspondence theorist might hold the right set of beliefs that will ensure coherence between a correspondence theory of truth and other propositions believed. *Given the coherence theory*, there seems no reason to insist that proponents of a correspondence theory of truth have a theory that is false (relative to their system of beliefs).

Sophisticated coherence theorists have learned to bite the bullet on this question. They concede that their version of antirealism applies to itself, and that one can, if one likes, think of truth as consisting in correspondence. One can think this provided that one realizes that the truth of the proposition that truth involves correspondence depends on the coherence of that proposition with other propositions believed.[3] It is almost irresistible to suppose, at this point, that the antirealist is really committed to the view that antirealism is true, period. You can pay lip service to any view you like provided that you understand that in the end the truth of your view can consist in nothing other than its coherence with other propositions to which you are committed. And if this is so, there is still an underlying inconsistency between the alleged commitment to a ''pure'' coherence theory of truth (one that applies to all propositions) and this implicit commitment to a nonrelativized truth about the relativization of truth. Rather than pursue the question of whether commitment to a coherence theory of truth involves a self-referential paradox, however, let us look more closely at how the proponent of such a view understands reference to the *relata* of coherence.

What makes *P* true is that it coheres with a set of other propositions *Q* that are believed. But a coherence theorist of truth does not have *facts* about beliefs in any but the trivial sense in which there are *truths* about beliefs. The coherence theorist thinks that the proponent of a correspondence theory is suffering fundamental confusion in thinking that one can invoke the concept of a fact to *explain* what makes a proposition true. To say that it is a fact that water has molecular structure H_2O *is just another way of saying* that it is true that water has molecular structure H_2O.[4] The expression ''the fact that *P*'' has precisely the same meaning as ''*P*'s being true,'' and it is an almost comical error, therefore, to suppose that one can think of a fact as a *truth-maker*. So if a coherence theorist says that what makes *P* true is *the fact* that it coheres with some set of propositions *believed*, we know that is just another way of saying that what makes *P* true is that it is true that *P* coheres with some set of propositions *Q* that is believed. But what makes it *true* that *Q* is believed? Presumably, there is but one answer available to a coherence theorist. What makes it true that *Q* is believed is that the proposition that *Q* is believed coheres with some set of propositions *R* that is believed. Now *R* is either the same as or different from the set of propositions *Q*. If *R* = *Q*, then we again must ask what makes it true that *Q* is believed. The account is becoming viciously circular. We have been told that the truth of the proposition that *Q* is believed consists in that proposition's cohering with the propositions *Q* that are believed, but we still need to be told what makes it *true* that *Q* is believed. If *R* is not identical with *Q*, then we need to understand what makes it *true* that *R* is believed, an answer that the theory could give only in terms of coherence with beliefs. But whatever beliefs the coherentist appeals to (the same or different), we will again need an answer to the question of what makes it *true* that those beliefs are held.

In chapter 2 we discussed a conceptual regress argument for the need to recognize a concept of noninferential justification. My suggestion is that we have once again encountered a vicious conceptual regress with a coherence theory of truth. Consider again the example of the philosopher who thinks that one can simply define goodness as instrumental goodness: What makes *X* good is always that *X* leads to something *Y* that is good. I argued that there is a vicious conceptual regress involved in such a view. The theory, in effect, never does, never *could*, finish telling us what makes *X* good. The concept of instrumental goodness, in order to be *intelligible*, must be grounded in an understanding of goodness that does not involve thinking of something else being good. The coherence theorist of truth never does, never *could*, tell us what makes it true that someone believes a proposition. The proponent of such a view can start talking, to be sure, but there will be no end to the talking. We never get a theory of truth. We never get the ground of truth. We never get an account we can understand of what makes it true that someone has a certain belief. Every possible account consistent with the theory must refer to the fact that the proposition that the belief is held coheres with other beliefs. But whatever beliefs are selected, we will need an account of what makes it *true* that *those* beliefs are held. If the coherentist keeps referring to the same set of beliefs, the vicious regress might best be construed as blatantly circular. If the coherentist tries to escape by invoking different beliefs each time the question is asked, the regress will be linear. Either way, no account of what makes it true that beliefs are held could be completed. But without an account of what makes it true that someone has a belief, we do not have an account of what makes any proposition true, because the theory attempts to understand the truth of other propositions in terms of the *truth* of propositions describing a coherence between a proposition and other propositions *believed.*

In explaining the objection, I presupposed one of the simpler "actual belief" coherence theories, that is, I presupposed that the proponent of such a view was defining truth in terms of coherence with other propositions actually believed. Would it make any difference if the coherence theorist of truth moved to one of the more sophisticated "hypothetical belief" models? Of course not. Remember, the difficulty has nothing to do with the problem of a finite mind having an infinite number of

beliefs. As we saw in chapter 2, there probably is no difficulty in supposing that people can have an infinite number of different beliefs. The
regress that faces the coherence theorist of truth is a conceptual regress.
We are not getting an account of truth because every time we try to
understand what makes one proposition true we are necessarily led to a
question about what makes yet another proposition true. It is far from
clear what coherence theorists have in mind when they employ counterfactuals having to do with the end result of an indefinitely long process
of inquiry. But fortunately for us, it does not matter what they have in
mind. They will inevitably face the vicious regress. There is no *fact* of
the matter for a coherence theorist—there are only truths about what
the end result of such an inquiry would be. When a coherence theorist
tells us that what makes *P* true is that it coheres with the set of propositions *Q* that would be believed at the end of an inquiry, we must ask
what would make it true that *Q* would be believed at the end of this
process of inquiry. And when we are told (as we must be told for the
theory to be consistent) that what makes it true that *Q* would be believed
at the end of this inquiry is that this proposition coheres with propositions *R* that would be believed at the end of this idealized inquiry, we
must ask again what makes it true that *R* would be believed at the end
of this inquiry, and so on *ad infinitum*.[5]
     The nice thing about this argument is that we know it will be effective
against *any* modification of a coherence theory of truth. To be a coherence theorist one must eventually start talking about the relata of coherence. But such talk presupposes the existence of the relata, that is to
say, presupposes that it is true that such relata exist. But as soon as the
existence of the relata is invoked, the coherence theorist will encounter
the vicious conceptual regress: that truth will presuppose an understanding of other truths which will presuppose an understanding of
other truths, and so on *ad infinitum*. Even if I were wrong, for example,
in supposing that a coherence theorist of truth must select the relevant
subclass of propositions with which true propositions must cohere to
those that are *believed*, there must be some other feature used to select
the propositions, and it will have to be true that the propositions in
question have that feature. The implicit reference to this other truth will
begin the vicious regress.
     The same problem, of course, will arise when the coherence theorist
invokes truths about the relevant relations of coherence obtaining. So
far we have worried only about how the coherence theorists can understand the existence of the relata. But of equal importance is the reference to the *fact* that the relevant relations of coherence hold between

the relata. Suppose we have an unproblematic understanding of *P* and some set of propositions *Q* that are (or would be) believed. What would make it true that *Q* is consistent with *P*, or that *Q* makes *P* probable? If we have a perfectly general coherence theory, there is again but one answer. What makes it true that *P* is consistent with *Q* is that the proposition that *P* is consistent with *Q* coheres with some other set of propositions *R*. But that coherence will itself involve consistency, and we need to ask what makes it true that consistency obtains between *R* and the proposition that *P* is consistent with *Q*. We have encountered yet another vicious conceptual regress. The theory that attempts to tell us what makes a proposition true, what grounds truth, never gives us something we can comprehend. We are led on a necessarily endless wild goose chase for the ultimate truth-maker.

Pure coherence theories of truth are literally unintelligible. They fall victim to a vicious conceptual regress. One might be able to circumvent the objection by turning to an impure coherence theory of truth. One might argue, for example, that one's theory of truth was never intended to apply to necessary truths, such propositions as that *P* is formally consistent with *Q*. And as we shall see later, it may not be wildly implausible to suppose that propositions asserting the existence of probability relations are themselves necessarily true. The introduction of just this impurity will not do the trick, however, since it is a contingent truth that people believe what they do, or would believe certain things after an inquiry. So our coherence theorist had better exclude truths about what people think, or believe, or represent, from the scope of the theory. But one should have very good reasons for introducing impurities into a theory of truth. We are going to want to know what *does* make it true that one proposition coheres with another, or that a person has certain beliefs. And we are not going to be happy with the coherence theorist if we get implicit commitment to brute facts as the truth-makers. If we have facts that are independent of their representations and that make true some propositions, why not suppose that there are such facts for all propositions? It is not as if the coherence theory of truth has independent plausibility. To be frank, it has always seemed to me absurd on the face of it. Even if coherence yields justification, it seems almost a truism that for many propositions (for example, contingent universal propositions) no amount of justification will ever entail the truth of the proposition, and that includes coherence of the proposition with other propositions that are or would be believed. To embrace a coherence theory of truth one would need to have deep metaphysical reservations about the very intelligibility of facts that are independent of their representation. If an impure coherence theory needs any such facts at all, it is hard to see what would continue to motivate one to abandon the much more natural conception of a world full of things and properties that is the ultimate subject matter of our assertions and beliefs and the only thing that can make true or false our beliefs.

Furthermore, it is likely that so-called impure coherence theories of truth will turn out not to be theories of *truth* at all. You will recall that a correspondence theorist will gladly recognize that the world contains facts about coherence, facts that can make true thoughts about coherence. If one is a correspondence theorist with respect to a set of truths X and one takes other propositions Y to be made true by coherence relations holding between the assertions contained in X, then surely the most natural interpretation of the view is that propositions in Y *assert* the existence of the relevant coherence relations. But again, if *that* is the view, our impure coherence theorist is just a correspondence theorist in disguise who takes the propositions in Y to be made true by facts—facts about coherence.

The fundamental objection to a coherence theory of truth can be summarized this way. The proponent of such a view implicitly takes certain truths as metaphysically unproblematic which, *given the theory*, its proponent has no excuse to presuppose. Facts about beliefs and relations of coherence are, to say the least, odd metaphysical atoms out of which to construct a world. There are wildly different views about what a belief is and equally divergent views about the nature of necessary connections. At least some views about belief, very popular today, hold that identity criteria for the content of a belief themselves involve complex causal relations that hold between various states of the world and the state that constitutes the belief. Ironically, this seems to be a view that has been held by some of the most prominent antirealists.[6] But any such view about the nature of belief merely exacerbates the regress problem discussed above. We are not permitted to appeal to representation-independent facts about a table as truth-makers for propositions describing the table. We must construct the truth of the table's existence out of relations holding between propositions believed and the proposition that the table exists. But the very existence of a belief, on the view we are discussing, itself presupposes relations with the very world that is constructed out of belief. So certain views about the nature of belief combined with a coherence theory of truth will generate a more superficial sort of vicious conceptual circle to go along with the more fundamental regress problem discussed earlier.

One of the reasons it is worth dwelling so long on a coherence theory of truth is that we can learn from its mistakes. At least some of the problems it faces are paralleled by problems facing a coherence theory of justification. The most fundamental objection to a coherence theory of truth is that it takes beliefs and coherence to be metaphysically unproblematic when it has no business doing so. One of the now familiar objections to a coherence theory of justification is that it takes beliefs and coherence to be epistemologically unproblematic when it has no business doing so. Let us turn now to the coherence theory of justification.

The Coherence Theory of Justification

Since a coherence theory of truth is unintelligible, we should consider only coherence theories of justification that presuppose a realistic, correspondence theory of truth. As I noted earlier, we can distinguish between pure and impure coherence theories of justification. A pure coherence theory of justification takes the justification of every belief to be a matter of coherence. An impure theory restricts the thesis to a subclass of justified beliefs. In the most influential recent work on coherence theories, for example, BonJour defends a coherence theory of justification for empirical beliefs only.

Like the coherence theorist of truth, the proponent of a coherence theory of justification must make clear the relations of coherence that are supposed to yield justification. Questions will also arise about the relata of coherence. This time, though, it is hardly controversial to claim that the relata of coherence that yield justification for a given belief will be other actual or hypothetical beliefs. There will again be questions, however, about which actual or hypothetical beliefs are relevant to determining whether or not a given belief is justified. The vast majority of philosophers who support a coherence theory of justification take the relevant beliefs to be those present in a single individual. What justifies *me* in believing *P* is that *P* coheres with some set of propositions that *I* occurrently or dispositionally believe or would believe were I to inquire or reflect in a certain way. What justifies *you* in believing *P* is *P*'s relation to propositions that *you* do or would believe. The theory, in other words, embraces what should be uncontroversial, namely that justification must be relativized to an individual. What justifies you in believing *P* need not be the same thing that justifies me in believing *P*, and indeed you can be justified in believing *P* while I am justified in believing *not-P*.

While justification relativized to an individual's belief system is the norm for coherence theories, one finds more and more interest in so-called social epistemology. Roughly the idea is that what justifies you in believing *P* is a matter not just of what you believe, but of what others in the community believe. A very crude social coherence theory of justification might hold that *S* is justified in believing *P* only if *P* coheres with the propositions believed by all or most members of the community. Because one can distinguish as many different communities as one likes, justification might again be relativized, but this time it will be relativized to a community. On one version of the view, the individual can be regarded as the limiting case of a community. So relative to my society, I might be justified in believing *P*. Relative to my religion within my society, I might be unjustified in believing *P*. And relative to the society of one that I constitute, I might be justified again in believing *P*. Virtually everything critical I have to say about the coherence theory that relativizes justification to the individual's belief system will apply mutatis mutandis to social coherence theories, and so I will focus my comments on what I take to be the simpler and more plausible of the coherence theories.

Let us consider first, then, a coherence theory that holds that *S* is justified in believing *P* at *t* only if *P* coheres with the rest of the propositions believed by *S* at *t*. Let us include among *S*'s beliefs both occurrent and dispositional beliefs. So far I have committed the coherence theorist to only a necessary condition for justification. For the moment let us leave open the question of whether the coherence theorist will be an access internalist or an access externalist, so that we may inquire further into the nature of coherence.

Coherence

In his excellent discussion of kinds of coherence (for use in a coherence theory of justification for *empirical* beliefs), BonJour points out that one is not going to get very far if coherence is understood merely in terms of consistency. Consistency is simply too easy to come by. If we imagine a person who believes eight different propositions completely unrelated, we will have imagined a person with a consistent belief system but we will hardly have imagined a person with a coherent belief system. If logical consistency is too weak a requirement to constitute the mainstay of coherence, Richard Foley has persuasively argued that it is also too strong a requirement.[7] By focusing on lottery-type situations, Foley thinks he can describe a set of inconsistent beliefs each of which is perfectly justified. If there are a thousand people in a lottery I know to be fair, I can justifiably believe of each participant that he or she will lose and also justifiably believe that not all of them will lose. If we label the participants *P1* through *P1000*, the following propositions cannot all be true: *P1* will lose, *P2* will lose, . . . , *P1000* will lose, and either *P1* will win or *P2* will win, or. . . , *P1000* will win. This is sometimes referred to as the lottery "paradox," but I agree with Foley that there is really no paradox at all. We simply do have perfectly good reason to believe each proposition even though the conjunction of these propositions is necessarily false. Of course we would not be justified in believing the conjunction, but it would be the fallacy of division to suppose that because we are unjustified in believing the conjunction we are unjustified in believing the conjuncts. There are many attempts to avoid the conclusion that one can have justified inconsistent beliefs, but they all either are implausible on the face of it or run the risk of inviting a fairly radical skepticism. The most common response to the suggestion that we can be justified in believing that *P1* will lose is that we can be justified in believing only that *P1* will probably lose. But are we to infer from this that whenever our justification for believing *P* leaves open the possibility of error, we should claim only to be justified in believing that *P* is probable? Since even common sense allows that there is at least a tiny probability that I have just gone mad and am suffering massive hallucinatory experiences, shall I give up all claims to be justified in believing propositions describing the familiar world around me? Is the skeptical conclusion to be won this easily?

After arguing persuasively for the conclusion that we can have justified inconsistent beliefs, Foley goes on to suggest that if we accept his conclusion we should give up on coherence theories of justification. Logical consistency, he argues, has always been a *minimal* requirement of coherence, and we have seen that logical consistency is not required for justification. Coherence, therefore, is not required for justification. This conclusion, however, might be a little too quick. The coherence theorist will almost certainly recognize that logical consistency is not enough for coherence, and in the light of Foley's objection the coherence theorist might consider abandoning logical consistency entirely to focus on probabilistic relations. I'll return to this suggestion shortly.

When BonJour notes that logical consistency is far too weak to constitute the core of coherence, he also cautions against requiring too much for coherence. Specifically, he warns against embracing the apparently very strong requirement of ideal coherence described by Blanshard (1939, 264): "Fully coherent knowledge would be knowledge in which every judgment entailed, and was entailed by, the rest of the system.'' Although the requirement of two-way entailment is obviously extraordinarily strong, it is interesting to note that it is very easy to have a system of beliefs in which each proposition believed is entailed by the rest of the propositions believed. Indeed, if we include dispositional beliefs, I can confidently claim to have a belief system in which each of my beliefs is entailed by the rest of what I believe. And the same is, or should be, true of everyone who has taken and remembers an introductory logic course. One of the truth-functional connectives we all learned was material implication. As long as we know its truth-functional definition, we know that if P is true and Q is true, then it is true that P materially implies Q and true that Q materially implies P. Consequently, I assume that if we believe P and believe Q, we will believe (at least dispositionally) that P materially implies Q and that Q materially implies P. But then for any two propositions P and Q that I happen to believe, there will be in my belief system propositions entailing each. P will be entailed by (Q and Q materially implies P), and Q will be entailed by (P and P materially implies Q). My belief system that appeared initially to be unconnected is in reality a system consisting of nice, neat logical entailments. BonJour worries about belief systems that are incoherent in spite of the fact that each proposition believed is entailed by the rest. But the source of his worry is that the belief system might be ''composed of two or more subsystems of beliefs, each internally connected by strong inference relations but none having any significant connection with the others'' (p. 97). In light of this point, we can conclude that while BonJour's hypothetical situation is formally possible, anyone who knows logic should have been able to remove the incoherence by adding the relevant beliefs about material implication. If the one subsystem consists of a set of propositions believed X, and the other subsystem consists of Y, then we can hook them up *deductively* simply by adding to our belief system, X iff Y.

Because entailment is so easy to come by, BonJour's own conditions for coherence (p. 98) are themselves far too easy to come by:

3. The coherence of a system of beliefs is increased by the presence of inferential connections between its component beliefs and increased in proportion to the number and strength of such connections.

4. The coherence of a system of beliefs is diminished to the extent to which it is divided into subsystems of beliefs which are relatively unconnected to each other by inferential connections.

Take any set of propositions believed, whatsoever. If one adds to the system the truth-functionally complex propositions entailed by the simple propositions, then each proposition will be entailed by the rest and there will be as many different deductive relations as one cares to develop. The irony is that deduction is usually thought of as the strongest sort of connection that can obtain between propositions in one's belief system. But, in fact, deductive relations are a dime a dozen once one learns the trick of how to manufacture at will a deductively valid argument for each of two propositions you believe.

The moral the coherence theorist might draw from the above observations is the importance of stressing *probability* relations as the glue of coherence. At the same time, one might also be cautious about considering entailment as just the limiting case of making probable. It may be that to have an ideally coherent system it is better to believe propositions that make probable each other where this relationship of making probable falls short of the kind of entailment that is often trivial, precisely because the proposition entailed was so obviously presupposed by the propositions that entail it.

Of course, if one stresses the importance of probability relations in one's account of coherence, it is incumbent on one to give a philosophical account of the relation of making probable. This is hardly a problem that faces only coherence theorists. Foundationalists who wish to avoid fairly radical skepticism need to move beyond their foundations to various commonsense truths made probable by the foundations. Foundationalists who are inferential internalists need to give an account of this relation of making probable that allows one epistemic access to truths of the form ''*P* makes probable *Q*.'' Therefore, we defer until chapter 7 a more detailed discussion of how the coherence theorist might understand the crucial probability relations that will at least partially constitute coherence.

Bad Arguments against Coherence Theories of Justification

Before we put the coherence theory of justification into the framework of the internalism/externalism debate and examine the fundamental problem the theory faces, it might be worth briefly commenting on some illegitimate objections to a coherence theory of justification. One such objection focuses on the extreme relativity of a coherence theory. You can have an internally coherent set of beliefs that justifies you in believing *P* while I can have an equally coherent set of beliefs that is inconsistent with yours and that justifies me in believing *not-P*. But is there any difficulty in holding a view that has this consequence? *Any* plausible theory of justification must allow that you can be justified in believing *P* while I am justified in believing *not-P.* On the classic foundationalist theories, for example, it might turn out that your experiences are radically different from mine where the difference allows you to legitimately infer radically different conclusions about the way the world is.

The concern with radical relativity is perhaps exacerbated by the vague feeling that a coherence theory of justification makes one's choice of what to believe far too *subjective.* I want to know what to believe and the coherence theorist tells me to come up with a coherent set of beliefs. But for every coherent set of beliefs I think of, I can think of another set inconsistent with the first but just as internally coherent. Is the decision as to what I should believe completely arbitrary and subjective?

Earlier, in addressing the question of whether the epistemic "ought" could be reduced to the "ought" of morality or prudence, we touched on the question of whether beliefs are events under our control. While it seems to me that it is logically possible for me to will a belief into existence, it also seems equally obvious that for the vast majority of my beliefs I just *find* myself with them. It is at best ingenuous to suggest that a coherence theory needs to tell people what to believe as if they were starting from scratch. If I had no beliefs except for belief in a coherence theory of justification, God knows how I would decide what to believe. The choice *would* be radically arbitrary. But in fact, when I try to decide whether or not I should believe some particular proposition, *P,* I do so against a backdrop of beliefs I already have. Still, could I not include *P* in my belief system by deciding to change enough of my other beliefs? Yes, if the "could" is understood in terms of logical possibility. No, if the "could" expresses causal possibility.

BonJour worries a great deal about the so-called input objection to coherence theories of justification. If justification is only a matter of getting one's beliefs to cohere, how can one accommodate the obvious fact that our belief systems must in some sense respond to our experience of the world? But again, it is not clear what the critic is worried about. Nothing in a coherence theory precludes the possibility of beliefs being *causally* affected by experience. Whether or not one can rationally believe that one's beliefs are *causally* affected by the world will always be a matter of whether or not one's belief in such causal connection coheres with other beliefs. But unless one assumes a coherence theory of truth, there is no reason to suppose that the *existence* of causal connections has anything to do with coherent beliefs. If one wants some stronger sense in which experience of the world yields justification, one is probably out of luck with a coherence theory. But one can hardly press this objection to a coherence theory without simply presupposing some version of foundationalism.[8]

There may be versions of the above arguments more sophisticated than those I have presented, but in the final analysis it seems to me that there is no real profit in pursuing these sorts of objections. Once we put the coherence theorist into the appropriate camp with respect to the internalism/externalism debate, we see that the coherentist faces far more fundamental problems in defending the view.

Coherence Theories of Justification and the Internalism/Externalism Debate

The contemporary internalism/externalism debate postdates many of the classic expositions of coherence theories, and many well-known coherence theorists seem almost oblivious to such questions as whether justified belief requires access by the believer to the relevant relations of coherence. To his credit, BonJour, in defending a coherence theory of empirical justification, addresses the internalism/externalism controversy head on. As you recall, BonJour defends the view that we called strong access internalism (for empirical justification). BonJour argues that an empirical belief can be justified by virtue of exemplifying some characteristic X only if the believer is aware that the belief has that characteristic and is aware that beliefs with that characteristic are likely to be true. It is a matter of considerable irony that BonJour used his strong access internalism to attack foundationalism. The source of the irony is that this same strong access internalism deals an equally fatal blow to his own coherence theory of justification. The coherence theorist identifies the characteristic X in virtue of which a belief is justified as its coherence with other beliefs. But a strong access internalist will deny that coherence alone can yield justification. For me to be justified in believing P I must be aware that (have at least a justified belief that) P coheres with the other propositions I believe. But to be aware that P coheres with my other beliefs I must be aware of contingent truths regarding what I do in fact believe. BonJour's coherence theory of justification entails that there is only one way to have a justified belief about what I believe. To be justified in believing that I believe some proposition Q, my belief that I believe Q must cohere with the other propositions R that I believe. But that will not be enough, for a strong access internalist will require that I be justified in believing that the proposition that I believe that Q does cohere with the other propositions R that I believe and that will require having a justified belief that I believe R. That justification will require awareness of still other beliefs, and so on ad infinitum. BonJour's coherence theory of justification faces the very same regress with which he charged traditional foundationalism. This should come as no surprise. I warned the reader in chapter 3 that it is folly to embrace strong access internalism. To understand what strong access internalism requires is to realize that it must involve a vicious regress.

It is a tribute to BonJour that he raises this problem against his own view. The trouble is that he does not solve it. He attempts to respond to the difficulty with his much discussed "doxastic presumption." Bon-Jour suggests, in effect, that to even engage in epistemological inquiry we must presuppose that we have at least an approximate grasp of our belief system. Let us not worry about whether a strong access internalist could really get by with only an approximate grasp of a belief system, and inquire into the status of the presumption.

The doxastic presumption is not itself a belief that can be justified, and BonJour seems indecisive on the question of whether it is simply inappropriate to seek justification for the presumption or whether it is just too bad for epistemologists that they are not going to get a justification for it. I suspect that the latter comes closest to BonJour's position. Indeed, if one reads carefully one finds BonJour conceding that his coherence theory of justification combined with his internalism entails extreme skepticism with respect to empirical justification: "What the discussion leading up to the Doxastic Presumption shows is precisely that a coherence theory of empirical justification cannot, in principle, answer this form of skepticism; and this seems to me to count in favor of the skeptic, not against him" (p. 105). "This form of skepticism" refers to skepticism with respect to the existence and content of one's beliefs. BonJour goes on to argue that "the failure to answer one version of skepticism does not in any way mean that there is no point in attempting to answer others" (p. 105). The idea seems to be that we may have to concede that we do not have any reason to believe that we have beliefs but that *if* we knew what we believed a coherence theory of justification would allow us to move full speed ahead in responding to the skeptical challenge. But it is not clear what the importance of the conditional is. If wishes were horses, beggars would ride. If BonJour has a theory of justification that *entails* that we do not have epistemic access to our beliefs, and such access is required in order to have justified beliefs about the world around us, then we do not have justified beliefs about the world around us. The matter is that simple.

If one can make any use of the doxastic presumption at all, it would be by way of a dialectical argument establishing that a certain form of skepticism has a kind of self-refuting character. We talked about self-refutation in chapter 2, distinguishing a number of different ways in which an argument can be viewed as self-refuting. BonJour might argue that the skeptic who claims that our beliefs are unjustified is at least implicitly claiming knowledge that we have these beliefs. The very practice of challenging the justificatory status of a belief presupposes access to beliefs. The skeptical conclusion itself is one believed by the skeptic, and in endorsing that conclusion the skeptic is again presupposing access to at least one belief. But it will be difficult to establish any formal self-refutation in the skeptic who argues that BonJour's coherence theory of justification cannot refute skepticism. Strictly speaking, a skeptic need not make any assertions about what people believe in putting forth a skeptical conclusion. The skepticism can be expressed conditionally: *if* someone believes that there is a God, an external world, other minds, a past, future regularities, and so on, then such a belief is unjustified. The conditional does not entail that anyone has beliefs. To be sure, the activity of putting forth skeptical arguments in response to the positions of other philosophers seems to presuppose that we know what people believe. It also seems to presuppose that there are other people, that there is a past, and that there is an external world. Skeptics respond to correspondence, think fondly of their childhood, and put their thoughts down on what they take to be paper. But as we noted in chapter 2, philosophical skepticism does not require that one purge oneself of beliefs for which one cannot find adequate philosophical justification. If one is going to use the fact that the skeptic acts *as if* certain beliefs were true as an excuse to reach the conclusion that such beliefs *need* no justification, we can save ourselves a lot of time and effort and simply take for granted the truths of common sense.

BonJour's coherence theory of justification was intended to cover only the kind of justification we have for empirical beliefs. In discussing his view, therefore, we strayed a bit from a pure coherence theory of justification. The pure coherence theory obviously faces even more difficulty when it comes to reconciling the theory with a strong access internalism. Such a view will require that the subject who has a justified belief has access not only to what is believed but also to the relations of coherence that hold between propositions believed. If it is a pure coherence theory, there is only one source of justification for the belief that *P* coheres with *Q* and that is coherence. But according to the coherentist who is a strong access internalist, to be justified in believing that *P* coheres with *Q* I must be justified in believing that the proposition that *P* coheres with *Q* coheres with the rest of my beliefs. But this would require being justified in believing that the proposition that *P*'s cohering with *Q* coheres with the rest of my beliefs, which cohere with the rest of my beliefs, and so on ad infinitum. We face not one but two vicious regresses if we try to combine our coherence theory with strong access internalism. This second regress *may* not be a problem for Bon-Jour provided that *none* of the relations that define coherence are contingent. Because probabilistic connections play such an important role for BonJour (and for any plausible coherence theory), he must specifically hold that at least some propositions of the form "*P* makes probable *Q*" are necessary truths knowable a priori. In fact I think this is his position, although I am not sure he realizes fully the implications of it. (For one thing, he could have saved himself a lot of time when it came to his crucial argument for the view that the coherence of a belief system makes likely the truth of the system. His position allows him to claim that it is a necessary truth that coherence makes likely truth, and one can know a priori that it does. Having explained his views about a priori knowledge he should have simply dropped the matter. Instead we get a long discussion about alternative *contingent* explanations for the fact that our belief systems are relatively stable.)

If we abandon a pure coherence theory of justification and restrict the relevance of coherence to, for example, belief in contingent truths, we can avoid one of the two vicious regresses. But the price we pay is steep. If we need, and can make sense of, noninferential justification for some beliefs—for example, beliefs in necessary truths—how can we take seriously global arguments against foundationalism? In particular, it seems obvious that we cannot take seriously BonJour's own argument against foundationalism. According to that argument, *no* property *X* of a belief could be a sufficient condition for the having of a noninferentially justified belief because the believer would also need access to the fact that *X* was present and access to the fact that the presence of *X* made likely truth. But if the argument is good at all, it is surely good for a priori justification. It is not completely clear what feature of a believer's situation is supposed to constitute the having of a priori justification or knowledge for BonJour, but whatever that feature is we can label it *X* and run his own argument against the claim that this does constitute knowledge or justification without inference.

Strong access internalism is fatal to a pure coherence theory of justification and to a coherence theory restricted to the justification of empirical beliefs. Although I illustrated the problems with respect to a coherence theory that relativizes justification to an individual's belief system, one can see that the problems are only exacerbated by a coherence theory that relativizes justification to a community's belief system. If it is impossible for the coherence theorist to give an account of our access to our own beliefs, you can be sure that it is not going to be any easier to give a coherentist account of access to the community's belief system.

One can, of course, introduce still further impurities into the view. One can recognize a class of beliefs that have a "privileged" place within one's belief system. One can avoid a regress with respect to getting knowledge of what we believe by simply giving oneself privileged access to one's representational states. If one gives oneself privileged access to one's intentional states, however, it is not clear why one should not include all mental states. And why not throw in some straightforward beliefs about the physical world around us?[9] But as more and more impurities enter into the theory, it is difficult to avoid the conclusion that the coherence theorists are simply foundationalists when it suits them. If a certain kind of belief has a privileged position, we need an account of what gives the belief this special epistemic status. If we get such an account, we will have a version of foundationalism. If we do not get such account, we should dismiss the view.

The problems faced by this internalist version of a coherence theory of justification parallel in a striking way the problems faced by the metaphysical coherence theory of truth. The proponent of a coherence theory of truth needs to assume an unproblematic ontological status for beliefs and relations of a coherence. But a coherence theory of truth has no business taking any facts to be ontologically unproblematic. Facts are truths and truths are to be understood in terms of coherence. The proponent of an internalist coherence theory of justification needs to assume an unproblematic epistemic status for beliefs about beliefs and relations of coherence. But an internalist coherence theory of justification has no business taking beliefs of this sort to be epistemically unproblematic. The metaphysical coherence theory generates a vicious conceptual regress. The epistemological coherence theory generates a vicious epistemic regress.

If we cannot reconcile a coherence theory of justification with strong access internalism, should the coherence theorist simply view strong access internalism as the villain and continue to embrace coherence as the essence of justified belief? Strong access internalism, I argued, is fatal to all theories of justified belief, and so this is hardly a move about which I would complain. I did argue, however, for a crucial distinction between strong access internalism and inferential internalism. The inferential internalist does hold that to be justified in inferring *P* from a set of propositions *E*, one must be justified in believing that *E* makes *P* probable. The coherence theory of justification is a theory according to which all justification is inferential. The pure theory holds that the only thing that can justify a belief is another belief. To avoid these difficulties, then, the coherence theorist will need to reject not only strong access internalism but also inferential internalism. Roughly, the coherence theorist (who relativizes justification to an individual's belief system) can maintain that *S*'s belief that *P* is justified provided that *P* does cohere with the rest of the propositions *S* occurrently or dispositionally believes *whether S knows it or not.*

Can one develop a plausible coherence theory of justification that is conjoined with inferential externalism? There are two main reasons why I think one cannot. The first is not really much of an argument. It is simply the observation that if one is going to be an inferential externalist one might as well embrace reliabilism or one of the other views that reduce the relevant epistemic connections to nomological connections. At least this seems more plausible to me *if* one's coherence theory of justification is unrelated to support for a coherence theory of truth. And since a coherence theory of truth is unintelligible, one had better divorce one's coherence theory of justification from a coherence theory of truth. The "nomological" versions of inferential externalism are surely more straightforward than the coherence theory. They can accommodate Foley's correct conclusions about the possibility of having justified inconsistent beliefs. They have a better (though not very good) chance of securing some sort of intimate connection between the having of justified beliefs and the having of true beliefs. If I were going to be an inferential externalist, I would not waste my time trying to develop the complex relations of coherence needed to make sense of a coherence theory of justification.

More significantly, however, the combination of inferential externalism and a coherence theory of justification seems to me even more vulnerable to counterexamples of the sort that BonJour raised against externalism. BonJour, you will recall, worried about people who had completely reliable clairvoyant powers but who had no reason whatsoever to suppose that they did. It seemed clear to him that the beliefs generated by these powers were irrational. Intuitions vary about this sort of case, and I suspect that one of the reasons that there is so much disagreement about what to say is that the hypothetical situation does presuppose a connection between the process that forms the belief and the truth of the belief. As I admitted in chapter 4, I suspect that the epistemic concepts analyzed by paradigm foundationalist forms of externalism *do* sometimes find expression in ordinary discourse. But the coherence theorist who is an inferential externalist does not even have the nomological connections to truth upon which to fall back. Suppose I believe twenty-eight very complex propositions. Suppose further that I reached these conclusions in an extraordinarily silly way. I was reading a book far too difficult for me and to amuse myself I decided to believe every fifth proposition I encountered. As it turns out, by a remarkable coincidence there is an extremely sophisticated proof that interrelates all these different propositions, a proof that only a handful of logicians in the world would be able to grasp. Is there any plausibility at all in holding that my beliefs are rational?

One might complain that I am trading on the ambiguity between justifiably believing a proposition and a belief's being justified. As we drew that distinction earlier, I am justifiably believing *P* only if there is a justification for me to believe *P and* the fact that I have this justification is causally sustaining my belief. As I described the situation, the cause of my beliefs had nothing to do with their coherence, and so we do not have to say that my beliefs are rational. At the same time the coherentist could insist that the existence of the coherence constitutes a justification for the beliefs. But in what sense does mere coherence constitute a justification for *me* to believe *P*? If the coherence is something I could not grasp if my life depended on it, in what sense is it there *for me*? One might just as well define justified beliefs as true beliefs and claim that the truth of what I believe constitutes a justification that is there for me even though it will result in my justifiably believing *P* only if it causally sustains the belief. Furthermore, with a little ingenuity we could make the existence of the coherence causally relevant to my believing the propositions that cohere. It may be that the author of the book chose to communicate this complex proof to a select group of people using a code. The code involved inserting the crucial propositions at five-proposition intervals. The coherence contributed to the author's awareness of the coherence, and the author's awareness of the coherence is a link in the causal chain that resulted in my believing them.

Inferential externalism is itself an implausible view when it comes to developing a concept of justification in which a philosopher might be interested. But when it is combined with a coherence theory of justification, it is wildly implausible. It fails to capture any epistemic concept that finds expression in any arena of discourse.

Notes

1. I appreciate some helpful comments Laurence BonJour has given me in connection with this and subsequent arguments. He commented on an APA paper I presented, "The Incoherence of Coherence Theories," a paper that has been worked into sections of this chapter. BonJour convinced me that historically important realism/antirealism debates (of the sort initiated by Bradley, for example) are best not construed as debates over a pure coherence theory of truth—see my subsequent distinction between pure and impure coherence theories of truth.

2. There are, of course, others which for reasons of space I am not discussing. These include the so-called redundancy theory. On this view, the expression "is true" is not a genuine predicate expression attributing some property or characteristic to the "bearer" of truth value. When one says "*P*" in the context of making an assertion, one adds nothing to the content of the assertion by saying instead "*P* is true." There is certainly something to the claim that the statements "The cat is on the mat" and "It is true that the cat is on the mat" *convey* the same information to the person with whom we are speaking. But the obvious explanation for this is that it is a convention of our language that when we *assert P* we are attributing to the proposition the characteristic of being true. It remains an obvious fact that to think of a thought and to think of a thought's being true are two quite distinct states of affairs.

3. See Goodman 1978 and Putnam 1978, especially part 4.

4. In this respect the coherence theory bears a family resemblance to the redundancy theory. The redundancy theorist claims that the assertions "*P*" and "*P* is true" have the same content. The coherence theorist thinks that reference to its being a fact that *P* is just another way of referring to its being true that *P*.

5. Again, if *R* is construed as identical to *Q* the account becomes viciously and blatantly circular.

6. Putnam, for example.

7. Foley 1979.

8. BonJour's (1985) discussion of the role observation plays in empirical justification is very complicated and potentially misleading. There are passages (pp. 116–18, for example) in which he appears to modify his coherence theory of empirical justification to require that there are at least some justified beliefs whose justification is owed in part to their having been caused in the appropriate way. At the crucial point, however, he remembers that he must recast any reference to actual causal connections in terms of subjective conceptions of such

connections (p. 123). When he returns to this thesis, he returns to a pure coherence theory of justification for empirical beliefs.

9. This is, of course, just what people like Quine seem to do when they try to find a special place within their coherence theory for observation statements. See, for example, Quine 1969, especially pp. 85–87.

Chapter Six

Externalism and Skepticism

We have looked at paradigm foundationalist versions of both internalism and externalism. I tried to develop what I take to be the most plausible representatives of the two views and along the way I offered some criticisms of externalism. My primary goal, however, was to discover the core of the internalism/externalism controversy. We found that there was not one but a number of controversies associated with the internalism/externalism debate. I think the two most fundamental issues separating paradigm internalists and externalists are the questions of whether fundamental epistemic concepts can be "naturalized," that is, reduced to nomological concepts, and the question of whether one takes access to inferential connections to be a necessary condition for inferential justification. I have introduced the labels "inferential internalism" and "inferential externalism" to refer to the two positions one might take on this last question.

We also briefly examined the coherence theory of justification in light of the internalism/externalism controversies. Because the coherence theory holds, in effect, that all justification is *inferential*, I suggested that we focus on the question of whether the coherence theorist should be an inferential internalist or an inferential externalist. I concluded that if the coherence theory of justification embraces inferential internalism, the view faces a fatal vicious regress. But a coherence theory of justification combined with inferential externalism is wildly implausible and vulnerable to obvious counterexamples.

In what follows I am primarily interested in the ways in which both internalist and externalist versions of foundationalism can respond to the traditional skeptical challenge. In this chapter I focus on the externalist's response to the classic skeptical arguments sketched in chapter 2. For convenience I focus primarily on reliabilism, but almost all of what I say will apply mutatis mutandis to other paradigmatic

159

externalists. My first aim is simply to be clear about the framework within which a foundationalist version of externalism will face the skeptical challenge. I want to understand where externalism leaves the philosopher when it comes to approaching normative epistemological issues in general, and these issues as they relate to skepticism in particular. But, as I implied earlier, I think the very examination of the way in which the philosophical externalist should approach skepticism may reveal the fundamental weakness of externalism as a metaepistemological account of concepts fundamental to *philosophical* concern with epistemology.

Externalism, Foundationalism, and the Traditional Skeptical Argument

In chapter 2 I tried to characterize what I take to be the fundamental structure of skeptical arguments. The skepticism I am most interested in is skepticism with respect to justified or rational belief. Furthermore, we are concerned, in the first place, with ''local'' rather than ''global'' skepticism. The skeptics we considered put forth arguments designed to establish that we have no justified beliefs with respect to certain *classes* of propositions. Thus we looked at skeptical arguments that concluded that we have no reason to believe propositions about the physical world, the past, other minds, the future, and so on. The traditional skeptic virtually always presupposed some version of foundationalism, presupposed that we do have noninferentially justified belief in at least some propositions. The presupposition was seldom stated explicitly, but one cannot read any of the important historical figures concerned with either advancing or refuting skepticism without reaching the conclusion that they took some propositions to be epistemically unproblematic, where their unproblematic character seemed to stem from the fact that one did not need to *infer* their truth from any other propositions believed. In both the rationalist and the empiricist tradition, at least some propositions about the content of one's current mental states were taken to have this unproblematic, noninferential character. I do not pretend that the British empiricists explicitly adopted an acquaintance theory of knowledge or justification for these uninferred but unproblematic truths, but I do think that it is the most plausible hypothesis as to what they were implicitly endorsing.

The first step, then, in advancing an argument for skepticism with respect to some kind of proposition is to establish that our access to the relevant truth is at best *indirect*. In the terminology we have developed, the skeptic begins by denying that we have noninferential knowledge, or noninferentially justified belief in the relevant sort of proposition. Thus skeptics with respect to the physical world deny that we have noninferential "direct" access to physical objects. The standard claim is that if we have justification for believing anything about the physical world, that justification reduces to what we can legitimately *infer* about the physical world from what we know about the character of our past and present sensations. The skeptic about the past claims that we have no direct—that is, noninferential—knowledge of the past. What we know or reasonably believe about the past is restricted to what we can legitimately infer about past events from what we know about the present state of our minds. The future is known to us only through what we know about the past and the present. Other minds are known to us only through what we know about the physical appearance and behavior of bodies. And the theoretical posits of physics are known to us only through what we can legitimately infer from the macroworld of familiar objects that can be observed (at least according to common sense) through the five senses.

One of the primary advantages that paradigmatic externalist accounts have in the battle against skepticism is the ease with which they can deny the crucial first premise of skeptical arguments. The class of noninferentially justified beliefs is likely to be much larger given an externalist epistemology. Notice that I say "likely" to be much larger. As far as I can see, virtually all externalist epistemologies entail that it is a purely contingent question as to which beliefs are justified noninferentially and which are not. On the reliabilist's view, for example, the question of whether or not one is noninferentially justified in believing at least some propositions about the physical world is a question about the nature of the processes that yield beliefs about the physical world and the nature of their "input." If we have been programmed through evolution to *react* to sensory stimuli with certain representations of the world, and we have been lucky enough to have "effective" programming, then we will have noninferentially justified beliefs about the physical world. If Nozick is right and our beliefs track facts about the physical world around us, and this tracking does not involve inference from other propositions, we will again have noninferentially justified beliefs about the physical world. If our beliefs about the physical world are acting like that reliable thermometer that Armstrong uses as his model for direct knowledge, if we are accurately registering the physical world around us with the appropriate representations, then again we have noninferential, direct knowledge of that world. Whether or not we have such noninferential justification for believing propositions describing the physical world, on any of these externalist ways of understanding noninferential justification, is a purely *contingent* matter.

That it is a contingent fact is not in itself surprising, nor is it a consequence peculiar to externalist epistemologies. It is certainly a contingent fact on the acquaintance theory that I am acquainted with the fact that I am in pain. It is a contingent fact that I am in pain and so obviously contingent that I am acquainted with it. It is less obvious on traditional foundationalisms that it is a contingent fact that we are *not* acquainted with certain facts. It might seem, for example, that one *could not* be acquainted with facts about the distant past, the future, or even the physical world if it is understood as a construct out of actual or possible experience or as the cause of certain actual and possible experience.[1] But even here it is difficult to claim that it is necessarily the case that conscious minds are not acquainted with such facts. There may be no God, but it is not obvious that the concept of a consciousness far greater than ours is unintelligible. If the concept of a specious present makes sense, such a consciousness may have the capacity to directly apprehend a much greater expanse of time than can finite minds. In any event, it is not *clear* that the class of facts with which *we* can be acquainted exhausts the facts with which all possible consciousness can be acquainted.

But even if the scope of noninferentially justified belief is contingent on both internalist and externalist versions of foundationalism, there are crucial differences. On traditional (internalist) versions of foundationalism, philosophers are at least in a position to address reasonably the question of the content of noninferentially justified belief. The philosopher is *competent*, at least as competent as anyone else, to address the questions of whether or not we have noninferentially justified beliefs in propositions about the physical world, for example. There are two sources of knowledge as to what we are noninferentially justified in believing. One is dialectical argument. The other is acquaintance itself. One can be directly acquainted with the fact that one is directly acquainted with certain facts.

On the classic externalist views, the facts that determine whether one is noninferentially justified in believing a proposition are complex nomological facts. Given paradigm externalism, it is not clear that a philosopher *qua philosopher* is even in a position to speculate intelligently on the question of whether or not we have noninferentially justified belief in any of the propositions under skeptical attack.[2] Because the externalist has reduced the question of what is noninferentially justified to questions about the nature of the causal interaction between stimuli and response, and particularly to the processes of the brain that operate on the stimuli so as to produce the response, the search for noninferential justification would seem to be as much in the purview of the neurophysiologist as the philosopher.[3] In the last two hundred years, the vast majority of philosophers simply have not had the training to do a decent job of investigating the hardware and software of the brain. But without this training, it hardly seems reasonable for philosophers to be speculating as to what is or is not a reliable belief-independent process. To be sure, some contemporary epistemologists are trying to "catch up" on developments in cognitive science and even neurophysiology, but I cannot help worrying that the experts in such fields will quite correctly regard these philosophers as simply dilettantes who, having tired of their a priori discipline, now want to get their hands dirty in the real-life work of science.

Given this possibility, it is ironic that so many philosophers find externalist analyses of epistemic concepts attractive precisely because they seem to capture the prephilosophical intuition that there *is* something direct about our knowledge of the physical world through sensations. Through sheer repetition of the arguments, many philosophers got used to talking about *inferring* the existence of a table from propositions about the character of sensation, or *inferring* propositions about the past from propositions describing present consciousness. But critics have correctly pointed out that if such claims are intended to be phenomenologically accurate descriptions of our epistemic relation to the world, they are hardly credible. Anyone who has tried to draw knows that it is very difficult to distinguish the world as it appears from the world as it is. That there is a conceptual distinction between phenomenological appearance and reality seems obvious. If the difficulty of artistic representation shows that we rarely reflect on appearances (as opposed to reality), it also shows that there is such a thing as appearance. A number of philosophers have argued that the most frequent use of "appears" terminology is not that of describing the phenomenological character of sensation, but rather that of expressing tentative belief.[4] When I say that he appears to be a doctor I am probably only indicating my tentative conclusion that he is a doctor.

But even if we recognize what Chisholm called the "epistemic" use of "appears," there is surely another use of the term that is designed to capture the intrinsic character of sensation. When I say that the people on the street below look like ants, I am not expressing the tentative conclusion that they are ants. Again, as Sellars pointed out, we cannot directly conclude from such examples that the descriptive use of "appears" gives us a "pure" description of experience uncontaminated by reference to the physical world. "Appears" sometimes has what Chisholm called a "comparative" use.[5] To say that X appears F in this sense is to say that X appears the way F things appear under some set of conditions. The people down below look like ants in the sense that they look something like the way ants look when you are relatively close to them under standard conditions. Such complex facts include reference to physical objects and their tendency to appear in certain ways under certain conditions, and consequently are implausible candidates for objects of direct acquaintance. But Sellars aside, it is difficult to avoid the conclusion that the comparative use of "appears" virtually presupposes some other way of understanding the phenomenological character of appearance. There is some way that things appear and it is that way of appearing that the artist must think about in trying to represent realistically some aspect of the world. But whether or not this "noncomparative"[6] use of appears exists and is intelligible, it does not alter the phenomenological fact that we do seldom, if ever, consciously infer propositions about the physical world from propositions describing the character of sensation.

We also seldom consciously infer propositions about the past from anything we might call a memory "experience." As I noted in chapter 2, the very existence of memory "experience" is far from obvious. And it is relatively seldom that our commonplace expectations about the future are formed as a result of careful consideration of premises describing past correlations of properties or states of affairs. When I expect my next drink of water to quench my thirst instead of killing me, I do *not* first consider past instances of water quenching thirst. It is useful to reflect carefully on this fact, for even most externalists will view this kind of knowledge as involving inductive *inference*. We must, therefore, be cautious in reaching conclusions about the role of phenomenology in determining whether a *justification* is inferential or not. To repeat a point made in chapter 2, we must distinguish questions about the causal origin of a belief from questions about the justification available for a belief.

We must also distinguish between occurrent and dispositional belief. It may be that I have all sorts of dispositional beliefs that are causally sustaining my beliefs when I am completely unaware of the causal role these dispositional beliefs play. In introducing this discussion I suggested that it was ironic that externalists would find attractive the fact that their externalism can accommodate the apparent phenomenological fact that far fewer commonsense beliefs involve inference than are postulated by traditional foundationalism. The irony is that phenomenology should have no particular role to play for the externalist in reaching conclusions about what is or is not inferentially justified. According to the externalist, the epistemic status of a belief is a function of the nomological relations that belief has to various features of the world. These nomological facts are complex and are typically not the kinds of facts that have traditionally been thought to be under the purview of phenomenology. I suppose an externalist can define some belief-producing process as "phenomenological." But again, even if one can describe such a process, it will be a contingent question as to what beliefs such a process might justify, a contingent question that goes far beyond the competency of most philosophers (and certainly most phenomenologists) to answer.

But perhaps I am being unfair in suggesting that the philosopher who is an externalist in epistemology has no particular *credentials* qualifying him to assess the question of whether the skeptic is right or wrong in denying the availability of noninferential justification for beliefs under skeptical attack. The skeptics, after all, had arguments in support of their conclusion that we have no noninferentially justified beliefs in propositions about the physical world, the past, the future, other minds, and so on. The externalists can at least refute those arguments based on their a priori reasonings about the correct metaepistemological position. The most common way of supporting the conclusion that we do not have noninferentially justified beliefs about the physical world is to point out that we can imagine someone having the very best justification possible for believing that there is a table, say, before him, when the table is not in fact there. A person who is vividly hallucinating a table can have just as good reason to think that the table exists as you do. But we can easily suppose that there is no table present before the victim of hallucination. If *direct* epistemic access to the table is anything like a real relation, then it cannot be present when the table is not present. But if the victim of hallucination does not have direct access to the table, and the victim of hallucination has the same kind of justification you have for thinking that the table exists (when you take yourself to be standing before a table in broad daylight), then you do not have direct access to the table either.

The reliabilist will deny the association between noninferential justification and direct access to the table. To have a noninferentially justified belief about the table's existence is to have a belief about the table produced by an unconditionally reliable belief-independent process.

The victim of hallucination has (or at least might have) a belief in the table's existence produced by an unconditionally reliable belief-independent process. It depends in part on how we define the relevant process. But if we think of the stimuli as something like sensations (which the hallucinator has), and the process as what goes on in the brain when sensation is assimilated and turned into representation, there is no reason why someone who is hallucinating cannot satisfy the conditions for having a noninferentially justified belief, assuming of course that the process in question really is unconditionally reliable. The reliabilist's metaepistemology allows at least a conditional response to the skeptic's attack. More precisely, the reliabilist can point out that a reliabilist metaepistemology entails that the skeptic's conclusion about the noninferential character of belief about the physical world does not follow. And, of course, everything the reliabilist says about the physical world applies to the past, other minds, and even the future. The reliabilist probably will not claim that beliefs about the future are noninferentially justified, but he should claim that there is no reason in principle why they could not be, and should continue to assert that the skeptic has no argument for the conclusion that we have no direct, that is, noninferentially justified, beliefs about the future.

Interestingly, not all externalists will reject the skeptic's claim about noninferential justification in the same way. Consider again the reliabilist's response to the argument from hallucination as a way of establishing that we have no noninferentially justified beliefs about the physical world. The crucial move for the reliabilist was to deny that we are forced to regard the hallucinatory situation as one in which the subject lacked a noninferentially justified belief. A causal theorist about direct knowledge, like Armstrong, might admit that in hallucinatory experience we lack noninferential knowledge, but continue to assert that in veridical experience we have such knowledge. This externalist is more likely to deny the skeptic's presupposition that we should say the same thing about the nature of the justification available to the victim of vivid hallucination and the person who has *qualitatively* indistinguishable veridical experiences. You will recall that there is one sense of "internalism" according to which the internalist holds that the conditions sufficient for justification are always states internal to the subject. If sensations are not themselves relations (a controversial claim, to be sure), and the sensory evidence of S and R is indistinguishable, and there is nothing else "inside their minds" to distinguish their epistemic state, then this internalist will insist that if the one has a certain kind of justification for believing something, then so does the other.

But a causal theorist thinks that the relevant question that determines the nature of the justification available for a belief involves the *origin* of the belief. The internal, that is, nonrelational, states of *S* and *R* can be qualitatively indistinguishable, but *S*'s internal states can result in *S*'s having a noninferentially justified belief by virtue of their being *produced* in the appropriate way. *R*'s internal states might bring about the very same belief, but because they were not caused by the appropriate facts they will not result in the having of a noninferentially justified belief. In short, the hallucinator's belief cannot be traced via sensation back to the fact about the world that would make the belief true. The person lucky enough to have veridical experience typically has a belief that can be traced back to the fact that makes the belief true. This is a perfectly clear distinction, and there is nothing to prevent an epistemologist from arguing that this just is the distinction that determines whether or not someone has a justified or rational belief. Furthermore, the question of whether the justification is inferential has only to do with the kinds of links in the causal chain leading to the relevant belief. If the causal connection goes directly from some fact about the physical world, to the occurrence of sensory states, to representations about the physical world, then there are no other *beliefs* that crucially enter the story. The justification that results will be justification that does not logically depend on the having of other justified beliefs.[7] It will be noninferential justification. So again, we can see how an externalist metaepistemology can put one in a position to claim that the skeptic has not established the crucial premise concerning the inferential character of our belief in the propositions under skeptical attack.

Even if externalism allows one to point out that the skeptic has not established the crucial first premise of the argument, it does not follow, of course, that the externalist has given any positive reason to suppose that the skeptic is wrong in claiming that the propositions under skeptical attack are not the objects of noninferentially justified belief. In chapter 2 I insisted that both skeptics and nonskeptics play on a level playing field. There is no "burden of proof" when it comes to fundamental issues in epistemology. If the philosopher wants to claim that we have noninferentially justified belief in certain propositions, then the philosopher can give us good reasons to think that such justification exists. The skeptic who wishes to deny that we have such justification can give us good reasons to think that it does not exist. The skeptic, however, also has a fall-back position. Without arguing that we have no noninferentially justified beliefs in propositions about the physical world, the past, other minds, and the future, the skeptic can move "up" a level and deny that we have any good reason to believe that we have noninferential justification for these beliefs. A strong access internalist can move from the proposition that we have no justification for believing that we have a noninferentially justified belief that P to the conclusion that we do not have a noninferentially justified belief that P. But the externalist rejects just such an inference. Even if we abandon strong access internalism, however, we might find skepticism that maintains that we have no justification for believing that we have a justified belief that P just as threatening as skepticism that concludes that we are unjustified in believing P. Before we consider the question of whether skepticism will arise at the next level up within an externalist epistemology, let us briefly discuss the externalist approach to normative issues involving inferential justification.

Skepticism, Externalism, and Inferential Justification

Most of the general observations made about the externalist's response to skeptical challenges concerning the class of noninferentially justified beliefs will apply as well to inferential justification. If the skeptic were to succeed in convincing the externalist that we are not noninferentially justified in believing propositions about the physical world, for example, the externalist presumably would argue that such beliefs are inferentially justified. The reliabilist, for example, would argue that if our beliefs about the external world result from input that includes beliefs about the internal and external conditions of perceiving, or even beliefs about the qualitative character of sensation, the relevant belief-dependent processes are conditionally reliable and therefore produce (inferentially) justified beliefs, *provided that the input beliefs are themselves justified.* The proviso is crucial, of course, and reminds us that to establish that first-level skepticism is false, the externalist who concedes that the justification is inferential in character must establish the existence of at least one unconditionally reliable process and at least one conditionally reliable process.

We noted in discussing the externalist's views about noninferentially justified belief that externalism has a potentially significant advantage in dealing with skepticism precisely because there are no restrictions on how large the class of noninferentially justified beliefs might be. As I indicated, there is no a priori reason for the externalist to deny even that we have noninferentially justified beliefs about the past and the future. Evolution might have taken care of us rather well when it comes to reaching true conclusions about the world, and evolution might have accomplished this end without burdening our brains with too many conditionally reliable belief-forming processes. Nozick's tracking relations can in principle hold between any fact and any belief, and the tracking relations *need* not involve any intermediate beliefs.

Just as the externalist's class of noninferentially justified beliefs can be very large in comparison to those recognized by traditional foundationalists, so the class of inferences recognized as legitimate by the externalist can be equally large. Consider again the reliabilist's position. There are no a priori restrictions on how many different kinds of conditionally reliable belief-dependent processes there might be. Valid deductive inference is presumably the paradigm of a conditionally reliable belief-dependent process. Classical enumerative induction may satisfy the requirements as well, provided that we find some suitably restricted characterization of the inductive ''process'' that succeeds in denoting and that takes care of grue/green riddles of induction.[8] I suspect most externalists will be reluctant to include perceptual beliefs among the beliefs produced by belief-dependent processes, but there is no reason why a reliabilist could not be a sense-datum theorist or an appearing theorist who holds that we do have at least dispositional beliefs about the qualitative character of sensation and who further holds that such beliefs are processed by conditionally reliable belief-dependent processes that churn out commonsense beliefs about the physical world. In short, take any kind of inference that people actually make and the reliabilist could hold that it involves a conditionally reliable belief-dependent process. All one needs to do is to formulate a description of the process that takes the beliefs one relies on as premises (the input) and produces the beliefs that constitute the conclusion (the output). The description will have to be such that we succeed in picking out a *kind* of process that does play the causal role described, but it will not need to involve any reference to the ''hardware'' of the brain. Indeed, we can try to *denote* the relevant process by directly referring only to the kind of premises and conclusion with which it is associated. Roughly, the idea is that we can try to denote a belief-dependent process *X*, for example, using the description ''the process (whatever it is) that takes premises like these and churns out conclusions like this.'' Of course, such a description is probably too vague to do the trick. The locution ''like these'' can hardly be said to characterize precisely enough a class of premises. One would need to characterize the relevant points of similarity to have a well-defined class of premises which could then enter into the definite description denoting the process that takes them as input.

If we consider any argument someone actually makes, there will be indefinitely many classes of propositions to which the premises and the conclusion belong, and that will enable us to formulate any number of different descriptions of belief-forming processes. This is *not* a difficulty for the reliabilist, for as long as we have a locution that succeeds in denoting a process playing a causal role, we can use conditionals to define the conditions under which it is or is not conditionally reliable. The fact that a single inference might be subsumed under a number of different reliable belief-dependent processes is hardly a problem. If the inference can be subsumed under the description of both a reliable and an unreliable process, the crucial question will be which process is causally determining the production of a belief. Thus, if someone trustworthy tells me today that it rained in New York, I can describe this as a case of processing testimony to reach a conclusion about the truth of what is testified to, or I can describe it as a case of taking a statement I hear involving the name ''New York'' and believing all of the noun clauses containing that name. The former, let us suppose, is a reliable belief-dependent process, whereas the latter is not. But you recall that in formulating descriptions of processes appealing to kinds of premises and conclusions, we are merely *hoping* to denote some process (presumably a complex brain process) that does take input and causally produce output beliefs. It does not follow, of course, that every definite description we formulate will succeed in denoting. In the hypothetical situation we are discussing, it may be that there is no programming in the brain that takes the ''New York'' input and processes it in the way described. If there is nothing denoted by the description playing the relevant causal role, then we do not need to worry about the fact that such a process, *if used*, would be unreliable. And we do not need to worry about the fact that we describe the inference in question *as if* it involved a belief-forming process of the sort described. If someone is programmed in such a way that she sometimes makes the legitimate ''testimony'' inference and sometimes makes the bizarre ''New York'' inference, the justificatory status of the resulting belief will be a function of the process that was causally operative in this case. If both processes are operating simultaneously, the reliabilist will probably need something like Nozick's conception of one method outweighing another. The justificatory status of the belief will depend on which belief-forming mechanism would have prevailed had they conflicted.

To emphasize the point made earlier, according to externalism there are indefinitely many candidates for legitimate inferential processes. There are no a priori restrictions on how many conditionally reliable belief-dependent processes might be operating in normal human beings. There are no a priori restrictions on how many belief-dependent tracking relations might exist between beliefs and the facts that they track. Furthermore, just as in the case of noninferential justification, the question of which inferential processes generate justified beliefs for the externalist will be a purely contingent fact of a sort inaccessible to most philosophers *qua* philosophers. The existence of conditionally reliable processes, tracking relations, and the like is something that could be discovered only as a result of empirical investigation into causal relations. Philosophers are not trained to engage in this sort of empirical investigation.

Externalism, Normative Epistemology, and the Limits of Philosophy

Based on the observations above, I argue that if externalist metaepistemologies are correct, then normative epistemology is an inappropriate subject matter for philosophy. Philosophers as they are presently trained have no special *philosophical* expertise enabling them to reach conclusions about which beliefs are or are not justified. Since the classic issues of skepticism fall under normative epistemology, it follows that if externalism were correct, philosophers should simply stop addressing the questions raised by the skeptic. The complex causal conditions that determine the presence or absence of justification for a belief are the subject matter of empirical investigations that would take the philosopher out of the easy chair and into the laboratory.

The realization that a good part of the history of epistemology becomes irrelevant to contemporary philosophy if we become metaepistemological externalists might cause a good many philosophers to reconsider externalism. I have always found the skeptical challenge to be fascinating and it has always seemed to me that I can address the relevant issues from my armchair (or my bed, depending on how lazy I happen to feel on a given day). If I had wanted to go mucking around in the brain trying to figure out the causal mechanisms that hook up various stimuli with belief, I would have gone into neurophysiology.

To rely on the philosopher's interest in skepticism and penchant for armchair philosophy as a rhetorical device to convert potential externalists, however, might be viewed as a new low in the art of philosophical persuasion. The mere fact that philosophers have been preoccupied with a certain sort of question does not mean that they were qualified to answer it. There are all kinds of perfectly respectable candidates for misguided philosophical investigations. Many philosophers, for example, have taken the question of whether every event has a cause to be a deep metaphysical issue in philosophy. As a good Humean, I would be the first to argue that it is a purely contingent question and if one wants to know the answer to it, one should not ask a philosopher. I am also sympathetic to the less popular view advanced by the positivists that metaethics exhausts the appropriate domain for philosophical investigation into morality. If a consequentialist analysis of right and wrong action is correct, for example, questions about what kinds of actions, or particular actions, we ought to perform are very complicated causal questions. Even if philosophy gives us some special insight into what is intrinsically good and bad (a thesis that is itself highly dubious), the question of which action would maximize that which is intrinsically good and minimize that which is intrinsically bad is the kind of question that philosophers are not particularly competent to address. The kind of person who is good at figuring out the consequences of actions is the kind of person who has extensive ''worldly'' experience and common sense. Perhaps philosophers of the past who were so preoccupied with normative issues in ethics had that kind of experience, but, without denigrating my profession, I humbly submit that today's academic is not the kind of person to whom one should turn for advice in dealing with real-world problems. The ivory towers of the philosophy professor are anathema to the kind of experience one needs to reach reasonable conclusions about what the world would look like if we behaved one way rather than another.

The question of whether normative ethics is a legitimate area of philosophy is far too controversial to settle with a few glib remarks. My only interest here is to point out that the history of philosophy need not constrain us when it comes to reaching conclusions about the appropriate subject matter of philosophy. Philosophers have worried about fundamental normative questions of ethics for well over two thousand years. That need not stop a philosopher from presenting a respectable argument for the conclusion that philosophical concern with ethics ends with the successful analysis of the subject matter of moral judgments. If one wants to do normative ethics in addition to metaethics, one will need to do the kind of empirical work that contemporary philosophers have not been trained to do. Analogously, the fact that philosophers have been preoccupied with the skeptical challenge for literally thousands of years should not stop contemporary epistemologists from entertaining the thesis that the appropriate subject matter of epistemology ends with metaepistemology. After the metaepistemological analysis is complete, the externalist might argue, the only way to answer normative questions in epistemology is to engage in the kind of empirical investigation that contemporary philosophers have not been trained *by philosophy* to do.

I suggested in some of my introductory comments that contemporary metaepistemological debate has the potential to change the very face of the philosophical study of epistemology. Specifically, I argue that if any of the paradigm externalist accounts of knowledge and justification are correct, we should dismiss all skeptical inquiry as irrelevant to the subject matter of philosophy. If the externalist is right, philosophers should stop doing normative epistemology.

In reaching this conclusion I should be careful to admit that the philosophical externalist can, of course, embed normative epistemological conclusions in the consequents of conditional assertions. One can talk about what one would be justified in believing were certain conditions to obtain. But these conditionals are still part of metaepistemology. Indeed, such conditionals are merely a way of illustrating the consequences of metaepistemological positions as they apply to particular hypothetical situations. A Nozick, for example, can discuss what one would or would not know about the external world *if* a tracking analysis of knowledge were correct and *if* our beliefs about the physical world track the facts that would make them true. Nozick's analysis of knowledge also has the interesting feature that we can apparently determine a priori that we do *not* know certain things, for example, that we do not know that there is no evil demon deceiving us. But there will be no positive normative claim with respect to empirical knowledge that Nozick is particularly competent to make *qua philosopher*. As we shall see in a moment, externalism does not prevent a philosopher from reaching rational conclusions about what one is justified in believing. My conclusion is only that a philosopher's philosophical expertise is nothing that helps in reaching such conclusions. To illustrate this claim more clearly, let us turn to the question of whether externalist metaepistemologies suggest that one should be a skeptic about whether or not one has justified belief.

Second-Level Skepticism and the Fundamental Problem with Externalism

It is tempting to think that externalist analyses of justified or rational belief and knowledge simply remove one *level* the traditional problems of skepticism. When one reads the well-known externalists, one is surely inclined to wonder why they are so sanguine about their supposition that our commonsense beliefs are, for the most part, justified, if not knowledge. When Nozick, for example, stresses that interesting feature of his account allowing us to conclude consistently that we know that we see the table even though we do not know that there is no demon deceiving us, we must surely wonder *why* he is so confident that the subjunctives that on his view are sufficient for knowledge are true. Perception, memory, and induction *may* be reliable processes in Goldman's sense, and thus given his metaepistemological position we *may* be justified in having the beliefs they produce, but, the skeptic can argue, we have no reason to believe that these processes *are* reliable, and thus, even if we accept reliabilism, we have no reason to conclude that the beliefs they produce are justified.

In the previous section I emphasized that if externalism is true then philosophers *qua philosophers* may not be particularly competent to answer normative questions in epistemology. I did *not* assert that if externalism is true we have no reason to believe that we have justified belief in commonsense truths about the world around us. According to externalist epistemologies, it is a purely contingent question as to what kinds of beliefs are justified. The existence of justified beliefs depends on nomological features of the world—facts about the reliability of belief-producing processes, the existence of tracking relations, causal connections between facts and beliefs, and the like. There are no a priori restrictions on what one might be justified in believing. But it *follows* from this that there are also *no* a priori restrictions on second-level knowledge or justified belief. It will also be a purely contingent question as to whether we have knowledge of knowledge or justified beliefs about justified beliefs. If we accept the externalist's metaepistemological views, it *may* be true that not only do we know what we think we know, but we also know that we know these things. Similarly, we may not only have all the justified beliefs we think we have, but we might also be justified in believing that we have these justified beliefs. The processes that yield beliefs about reliable processes may themselves be reliable. The beliefs about the truth of the subjunctives that Nozick uses to define first-level knowledge might themselves be embedded in true subjunctive conditionals that, *given the metaepistemological view*, are sufficient for second-level knowledge. My belief that my belief that *P* tracks the fact that *P* might track the fact that my belief that *P* tracks the fact that *P*. And there is no greater problem in principle when we move up levels. A reliable process might produce a belief that a reliable process produced the belief that my belief that *P* was produced by a reliable process. There might be a tracking relation tracking the tracking relation that tracks the fact that my belief that *P* tracks the fact that *P*. To be sure, the sentences describing the conditions for higher levels of metajustification might look more like tongue-twisters than metaepistemological analyses but, as ugly as they are, they are perfectly intelligible, and there is no a priori reason why the conditions required for higher-level justified belief and knowledge might not be satisfied.

It is also important to note that according to the externalist, in order to be justified in believing that I have a justified belief that *P*, I need not know anything about the *details* of the nomological connections sufficient for knowledge. Consider again reliabilism. In order to be justified in believing that my belief that *P* is produced by a reliable process, I do not need to know the physiological details of the brain states linking stimuli and belief. I would need to believe that there is *some* process producing the belief and I would need to believe that the process is reliable, but I would not need to know very much about what that process is. As I indicated earlier, one can denote the processes that produce beliefs using definite descriptions that refer directly only to the kinds of premises and conclusions that are linked by the process. Of course, the definite descriptions might fail to denote, and the beliefs in propositions expressed using such definite descriptions will either be false or meaningless (depending on what one does with the truth value of statements containing definite descriptions that fail to denote). But the descriptions might be successful, and in any event the belief that there is a reliable process taking stimuli *S* and resulting in belief *P* might itself be produced by a reliable process.

All this talk about what would in principle be possible given an externalist metaepistemology is fine, the skeptic might argue. But *how* exactly would one justify one's belief that, say, perception and memory are reliable processes? The rather startling and, I think, disconcerting answer is that *if* reliabilism is true, and *if* perception happens to be reliable, we could *perceive* various facts about our sense organs and the way in which they respond to the external world. Again, *if* reliabilism is true, and *if* memory is reliable, we could use memory, in part, to justify our belief that memory is reliable. You want a solution to the problem of induction? There is potentially no difficulty for the externalist. If reliabilism is true, and if inductive inference is a conditionally reliable belief-dependent process, then we can inductively justify the reliability of inductive inference. Our inductive justification for the reliability of inductive inference might itself be reliable, and if it is, that will give us second-level justification that our inductive conclusions are justified. A solution to the problem of induction will be important because with induction giving us inferentially justified conclusions, we can use inductive inference with the deliverances of perception and memory to justify our belief that those processes are reliable. I can remember, for example, that I remembered putting my keys on the desk and I can remember the keys being on the desk. If memory is an unconditionally reliable belief-independent process, then both my belief that I remembered putting the keys on the desk and my belief that I put the keys on the desk will be justified. I now have a premise that can be used as part of an inductive justification for memory being reliable. The more occasions on which I can remember memory being reliable, the stronger my inductive argument will be for the general reliability of memory.

The skeptic could not figure out how to get from sensations to the physical world. Assume that perception is itself a belief-independent, unconditionally reliable process. Assume also that whatever perception involves, its specification involves reference to sensation, and assume further that we have "introspective" access to sensation. Introspective access might itself be another belief-independent, unconditionally reliable process. Given these suppositions, if reliabilism is true, then introspection can give us justified beliefs that we are perceiving, and perception can give us justified beliefs that physical objects are present. The two reliable processes together can furnish a premise that, when combined with others generated in a similar fashion, gives us inductive justification for believing that perception is reliable. So if both introspection and perception happen to be reliable, there seems to be no great obstacle to obtaining justified belief that they are reliable. Second-level justified belief is not much more difficult to get than first-level justified belief.

How successful *inductive* reasoning will be in answering second-level skeptical questions depends very much on how the externalist resolves some of the controversies discussed in chapter 4, specifically on how narrowly the relevant belief-forming processes are characterized. I have pointed out that as long as reliability is not defined in terms of actual frequencies, there is no *conceptual* difficulty in a reliabilist positing the existence of very narrowly defined, reliable belief-forming processes that have only a few, or even no, instances. Although there is no conceptual difficulty in supposing that there are such processes, it obviously creates problems for any *inductive* justification for believing that they exist and are reliable. As should be clear by now, however, the unavailability of inductive justification in no way implies that there is not some *other* reliable belief-forming process that will still yield second-level knowledge or justified belief.

This reminds us, of course, of Quine's injunction to naturalize episte-mology.[9] Quine suggested that we give ourselves full access to the deliverances of science when it comes to understanding how we have knowledge of the world around us. Contemporary externalists have simply given us more detailed metaepistemological views which allow us to rationalize following the injunction to naturalize epistemology. If the mere reliability of a process, for example, is sufficient to give us justified belief, then *if* that process is reliable we can use it to get justified belief wherever and whenever we like.

All of this will, of course, drive the skeptic crazy. You cannot *use* perception to justify the reliability of perception! You cannot *use* memory to justify the reliability of memory! You cannot *use* induction to justify the reliability of induction! Such attempts to respond to the skeptic's concerns involve blatant, indeed pathetic, circularity. Frankly, this does seem right to *me* and I hope it seems right to *you*, but *if* it does, then I suggest that you have a powerful reason to conclude that externalism is false. I suggest that, ironically, the very ease with which externalists can deal with the skeptical challenge at the next level betrays the ultimate implausibility of externalism as an attempt to explicate concepts that are of *philosophical* interest. If a philosopher starts wondering about the reliability of astrological inference, the philosopher will not allow the astrologer to read in the stars the reliability of astrology. Even if astrological inferences happen to be reliable, the astrologer is missing the point of a *philosophical* inquiry into the justifiability of astrological inference if the inquiry is answered using the techniques of astrology. The problem is perhaps most acute if one thinks about first-person philosophical reflection about justification. If I really am interested in knowing whether astrological inference is legitimate, if I have the kind of philosophical curiosity that leads me to raise this question in the first place, I will not for a moment suppose that further use of astrology might help me find the answer to my question. Similarly, if as a philosopher I start wondering whether perceptual beliefs are accurate reflections of the way the world really is, I would not dream of using perception to resolve my doubt. Even if there is some sense in which the reliable process of perception might yield justified beliefs about the reliability of perception, the use of perception could never satisfy a *philosophical curiosity* about the legitimacy of perceptual beliefs. When the philosopher wants an answer to the question of whether memory gives us justified beliefs about the past, that answer cannot possibly be provided by memory.

Again, if one raises skeptical concerns understanding fundamental epistemic concepts as the externalist does, then there should be no objection to perceptual justifications of perception, inductive justifications of induction, and reliance on memory to justify the use of memory. If one is understanding epistemic concepts as the reliabilist suggests, for example, then one can have no objection in principle to the use of a process to justify its use. After all, the whole point of inferential externalism is to deny the necessity of having access to the probabilistic relationship between premises and conclusion in order to have an inferentially justified belief. The mere reliability of the process is sufficient to generate justified belief in the conclusion of an argument. There is no conceptual basis for the reliabilist to get cold feet when epistemological questions are raised the next level up. Either reliability alone is sufficient or it is not. If it is, then it is sufficient whether one is talking about justification for believing *P* or justification for believing that one has a justified belief that *P*.

It is both interesting and illuminating that even many access externalists seem to worry about the possibility of second-level justification in ways that they do not worry about the possibility of first-level justification. As we noted earlier, Alston explicitly rejects the idea that one needs access to the adequacy of one's grounds for believing *P* in order to be justified in believing *P*. But in *The Reliability of Sense Perception*, he also seems to reject the idea that one can use a ''track record'' argument (an inductive argument of the sort I sketched above) to *justify* one's belief that perception and memory are reliable. Such arguments will inevitably presuppose the adequacy of the very grounds whose adequacy is at issue. In doing so the argument will be viciously circular.

But what exactly is Alston's complaint? To justify my belief that perception or memory is reliable, I need only find a good argument whose premises I justifiably accept and whose premises support the conclusion that these ways of forming beliefs are reliable. But if perception and memory are reliable *and there is no requirement of access to adequacy of grounds* in order for a belief to be justified, what is the problem? Why is it harder to justify my belief that my perceptual beliefs are justified than it is to have justified beliefs based on perception?

Much of the time Alston seems to admit everything I have just said, but the circular nature of the available arguments still clearly bothers him:

if sense perception is reliable [Alston's emphasis], a track record argument will suffice to show that it is. Epistemic circularity does not in and of itself disqualify the argument. But even granting that point, the argument will not do its job unless we *are* justified in accepting its premises; and this is the case only if sense perception is in fact reliable.

. . . But when we ask whether one or another source of belief is reliable, we are interested in *discriminating* those that can reasonably be trusted from those that cannot. Hence merely showing that if a given source is reliable it can be shown by its record to be reliable, does nothing to indicate that the source belongs with the sheep rather than with the goats. (p. 17)

But again, *as an externalist*, what does Alston want? He obviously thinks that in some sense all we could ever really conclude is that we *might* have justification for thinking that we have justified beliefs based on perception. And the contextual implication of this claim is that we also *might not*. But what "might" is this? Clearly, it is intended to refer to *epistemic* possibility. Let us say that *P* is epistemically possible for *S* when *P* is consistent with everything that *S* knows. Is it epistemically possible for us that perception is unreliable? Not if perception is reliable, because we will have inductive knowledge that it is not unreliable. But you are still just asserting a conditional, Alston will complain. For all we know, it is possible that perception is unreliable. But this claim about epistemic possibility is precisely the claim that Alston, as an externalist, has no business making. Can we *discriminate* (his word) between reliable and unreliable fundamental sources of belief? As an externalist he has no reason to deny that we can and do discriminate between reliable and unreliable processes (using, of course, reliable processes). Alston clearly wants to assert (and assert justifiably) a conclusion about epistemic possibility. But the concept of epistemic possibility he wants to apply at the second level is not one that can be understood within the framework of the externalism he embraces.

I agree, of course, with Alston's conclusion that one cannot use perception to justify one's belief that perception is reliable and memory to justify one's belief that memory is reliable. But that is only because the externalist is wrong in characterizing the concept of justification that even externalists are often interested in when they move up levels and start worrying about whether they can justify their belief that their beliefs are justified. The epistemic concept of discrimination that Alston invokes in the passage I quoted is precisely the concept that is at odds with his own attempt to defend an externalist understanding of epistemic concepts.

The fundamental objection to externalism can be easily summarized. If we understand epistemic concepts as the externalists suggest we do, then there would be no objection in principle to using perception to justify reliance on perception, memory to justify reliance on memory, and induction to justify reliance on induction. But there is no philosophically interesting concept of justification or knowledge that would allow us to use a kind of reasoning to justify the legitimacy of using that reasoning. Therefore, the externalist has failed to analyze a philosophically interesting concept of justification or knowledge.

The objection is by no means decisive. Obviously, many externalists will bite the bullet and happily embrace Quine's recommendation to naturalize epistemology. If the argument convinces anyone, it will be those who were initially inclined to suppose that externalism will inevitably encounter skepticism at the next level up. Maybe we have knowledge or justified belief as the externalist understands these concepts, some would argue, but we would never be in a position to know that we have knowledge or justified belief if the externalist is right. The only reason I can see for granting the first possibility but denying the second is that one is implicitly abandoning an externalist analysis of epistemic concepts as one moves to questions about knowledge or justification at the next level. But if when one gets philosophically "serious" one abandons the externalist's understanding of epistemic concepts, then, for philosophical purposes, one should not concede the externalist's understanding of epistemic concepts at the first level. Once you concede that according to the externalist we might have knowledge or justified belief about the past and the external world, you have also implicitly conceded that we might have knowledge that we have such knowledge, justified belief that we have such justified belief. And we might also have knowledge that we have knowledge that we have knowledge, and have justified beliefs that we have justified beliefs that we have justified beliefs. It seems to many of us that the externalist is simply missing the point of the philosophical inquiry when externalist analyses of epistemic concepts continue to be presupposed as the skeptical challenge is repeated at the metalevels. But the only explanation for this is that the externalist analysis of epistemic concepts never was adequate to a philosophical understanding of epistemic concepts.

Notes

1. For a detailed defense of this last view, see Fumerton 1985.
2. I stress "qua philosopher" for there is a real danger that I will be misun-

derstood on this point. Later in this chapter I argue that externalism is perfectly compatible with philosophers (and anyone else) having justified beliefs about whether or not they have justified beliefs. It will not, however, be their philosophical competence that yields such justification.

3. As I shall argue shortly, this claim might be misleading. In one sense the detailed character of belief-forming processes would be best discovered by neurophysiologists. But there is another sense in which anyone can form beliefs about such processes, even without any detailed knowledge of how the brain works.

4. See Sellars 1963, pp. 146–47.

5. Chisholm 1957, p. 49.

6. Again the terminology and the distinction is introduced by Chisholm (1957, 50–53).

7. It should go without saying that there may be causally necessary conditions for the existence of such noninferential justification having to do with the capacity to form other beliefs. The dependency that concerns us, however, is logical. Justification is noninferential when no other belief is a *constituent* of the justification.

8. The allusion is, of course, to the problem discussed in Goodman 1955, chap. 3.

9. Quine 1969, chap. 3.

Chapter Seven

Internalism and Skepticism

If one accepts an externalist metaepistemology, the philosophical battle with skepticism is over. It is not that the externalist wins the battle. Rather, the philosophical externalist becomes a conscientious objector. The externalist's account of what would make a belief epistemically rational implies that a philosopher qua philosopher is incompetent to evaluate skeptical claims. If externalism is true, the skeptic may be wrong. We may also have good epistemic reason to think that the skeptic is wrong. We may have good epistemic reason to think that we have good epistemic reason to think that the skeptic is wrong. But *philosophical* reflection and expertise will not get us these good epistemic reasons, nor will it put us in a position to recognize them if they exist. The most the externalist can reasonably point out is that if externalism is true, then the skeptic has given us no good argument *for* skepticism. And perhaps that is all the externalist wants—a refutation of skeptical arguments rather than a refutation of skeptical conclusions.

If externalism makes it easy to dismiss the skeptic's *arguments*, the internalist faces an uphill battle in the attempt to avoid skeptical conclusions. As we saw in chapter 2, the principle of inferential justification, which plays such an important role in the skeptic's reasoning, is built into the inferential internalist's *analysis* of what justification is. The internalist will not be able to reject skepticism by rejecting this presupposition of traditional skeptical arguments. In the following evaluation of internalist responses to skepticism, I am most interested in making *clear* the dialectical alternatives. The internalism I recommend, the inferential internalism based on the sui generis concept of acquaintance, has painfully few options when it comes to dealing with the skeptical challenge, and once we realize what they are, we can see that the view offers us limited but clear choices.

Internalism, Noninferential Justification,
and the Traditional Skeptical Argument

In the last chapter we saw that the externalist can easily deflect the first step in skeptical arguments for local skepticism. The skeptic with respect to some class of propositions, *S*, you recall, typically begins by denying foundational status to belief in this sort of proposition. The claim is made that we never have *noninferential* justification for believing propositions of this sort. The argument for this premise involves a thought experiment in which we are to compare veridical and nonveridical experience. In performing the thought experiment we are supposed to see that the nature of the justification is the same in both cases, that in the nonveridical situation it is implausible to conclude that the justification is noninferential, and thus that it is unreasonable to conclude of the veridical situation that it could be noninferential. To illustrate with skeptical arguments concerning perception, a vivid hallucinatory experience can give me precisely the same justification for thinking that there is a table before me (when there is not) that sensation would give me were I veridically perceiving a table. In hallucinatory experience, it is implausible to claim that I have any kind of *direct*, unproblematic access to the table. But since the nature of the justification is the same in both cases, one cannot insist that veridical experience gives one *direct* and unproblematic access to a table.

The argument is ineffective against externalists. Because the existence of noninferential justification is a function of the causal origin of a belief, appeal to the phenomenological indistinguishability of two hypothetical situations is neither here nor there when it comes to the question of whether a believer has the same or different justification. Furthermore, a reliabilist understands noninferential justification in such a way that the absence of a table does not imply that one lacks noninferential justification for believing that the table exists. Externalists probably will not talk about *direct* access to the fact that the table exists. Nevertheless, when we are hallucinating and believe that there is a table before us, we might have noninferential justification that derives from the fact that the belief is produced by a belief-independent reliable process. This externalist admits that in both hallucinatory and veridical experience we might have precisely the same justification for believing that the table exists, but insists that there is no reason to abandon the claim that in *both* situations the belief in question is noninferentially justified.

Does the first step of the traditional skeptical argument fare any better when it comes to internalism? It does. At least it does if we adopt what I take to be the most plausible version of foundationalism, the version that insists that the paradigm of noninferential justification is constituted by a direct relation that obtains between a believer and the *fact* that makes true the belief. If we are hallucinating a table that does not exist, then our justification for believing that the table exists cannot consist in our being directly and immediately acquainted with the fact that the table exists. By hypothesis, there is no fact of the relevant sort there with which we can be acquainted. Our attention must then switch to the claim that our justification for believing in the existence of the table is the same in veridical experience as it is in vivid hallucinatory experience.[1] A crude causal theory of knowledge, you recall, can no more hold that hallucinatory experience gives one direct knowledge of the table than can the acquaintance theory of direct knowledge. The difference is that the acquaintance theorist will probably acknowledge that acquaintance itself is a source of knowledge as to the character of justification. One can be acquainted with the fact that one is acquainted with a fact, and one can be acquainted with the fact that one's relation to the world is the same in two hypothetical situations.

The reader might worry at this point that I am backing away from my earlier rejection of weak access requirements for justification and knowledge as the fundamental defining distinction between internalists and externalists. It is true that I do not think that one's views about access requirements, strong or weak, lie at the *heart* of the internalist/ externalist debate. It is too easy for externalists to mimic access requirements from within their externalist framework. But I also admitted that most internalists will allow that when one is noninferentially justified in believing *P*, one *can* be noninferentially justified in believing that one is noninferentially justified in believing *P*. And the possibility in question is not merely logical or lawful. One's present epistemic situation typically puts such access easily within one's grasp. I can access my noninferential justification in the same sense in which I can remember what I had for breakfast an hour ago. The possibility of such access to noninferential justification is *not*, I hasten to remind you, a *constituent* of noninferential justification. It is not a defining necessary condition for noninferential justification. I observe only that most internalists will allow that one does in fact have noninferential access to noninferential justification, at least at the first level of beliefs. (As I pointed out earlier, when we move up levels, the sheer complexity of the propositions one must entertain may preclude forming the beliefs necessary to have noninferentially justified beliefs about the epistemic status of the beliefs below.) In particular, I think that I am directly acquainted with similarities and differences with respect to the relations of acquaintance in which I stand. If I can reason that there would be nothing to reveal a distinction between what I am acquainted with in hallucinatory and veridical experience, and I can reason that I am not directly acquainted with facts about the physical world in hallucinatory experience, then I can conclude that I am not directly acquainted with such facts in veridical experience.

I have illustrated the first step in the skeptic's argument by looking at local skepticism with respect to the physical world. When we examined the pattern of skeptical arguments in chapter 2, we saw that essentially the same first step is made in advancing local skepticism with respect to the past, the future, other minds, and theoretical entities in science. If we consider the present state of consciousness that constitutes seeming to remember having done *P*, we can again distinguish veridical and nonveridical memory. In nonveridical memory we are not directly acquainted with a past event (because the event did not even exist), and so in veridical memory it is equally implausible to hold that we are directly acquainted with a past event. I am after all acquainted with the similarity of the two epistemic situations.

It becomes more awkward to use the locutions ''veridical'' and ''nonveridical'' to describe the contrasting experiences that yield true and false beliefs with respect to the future and other minds. But terminological difficulty aside, it is clear that we can contrast two phenomenologically identical situations in which one has true and false beliefs, respectively, with respect to propositions about the future or other minds. In the situations involving false belief there was no fact with which we could be related that would yield noninferential justification, and thus the justification available in the phenomenologically indistinguishable situation involving true belief cannot be identified with direct access to the relevant part of the world.

There is one qualification I must add. Earlier I suggested that even on an acquaintance theory one might want to allow the possibility of noninferentially justified false beliefs. It seemed to me that acquaintance with a fact very similar to the fact that would have made *P* true might noninferentially justify me in believing that *P* (when I have the thought that *P* and am acquainted with a relation very much like correspondence holding between the thought that *P* and the spurious fact that *P*). If one adopts this suggestion, it is less obvious that in hallucinatory experience of a table, for example, one lacks noninferential justification for believing that the table exists. One could, in principle, argue that even though in hallucinatory experience the table was not there, there was a fact very much like a table's existing, and my acquaintance with that fact gives me a noninferentially justified belief in the table's existence.

The obvious reply to this argument, however, is to point out that there is nothing in hallucinatory experience at all like the fact that a table exists. To be sure, there are constituents of hallucinatory experience qualitatively identical to constituents of veridical experience. Indeed, it is precisely this conclusion that generates arguments for sense data or ways of being appeared to. But the common constituents of both hallucinatory experience and veridical experience do not seem to be similar facts about the physical world. There is nothing like a table's existing that is present in hallucinatory experience *unless* one introduces into one's ontology states of affairs that do not obtain. With nonoccurrent or nonactual states of affairs, I suppose one could understand the difference between hallucination and veridical experience in the following way. In hallucination, one is acquainted with the nonoccurrent state of affairs, the table's existing. In veridical experience, one is acquainted with the occurrent state of affairs, the table's existing. Now, to be honest, I have never really understood such views. I cannot figure out what this property of being occurrent is that some states of affairs have and some lack. But whatever the property is supposed to be, I assume that it makes a huge difference in the character of the state of affairs and that it should be difficult to confuse a nonoccurrent state of affairs with which one is acquainted, with an occurrent state of affairs with which one is acquainted. If so, it would again be implausible on an acquaintance theory to construe hallucinatory experience as yielding noninferentially justified belief in propositions about the physical world, justification that derives from acquaintance with spurious physical facts easily confused with real facts about the physical world.

Skepticism and Inferential Internalism

If the inferential internalist accepts the first move in the skeptic's argument, the refutation of skepticism becomes either easy or impossible, depending on how one answers certain fundamental questions concerning the concept of probability relevant to epistemology.[2] Once the internalist accepts the premise that we must be inferentially justified in believing propositions about the physical world based on what we know about sensation, inferentially justified in believing propositions about the past based on what we know about present consciousness, inferentially justified in believing propositions describing the mental states of others based on what we know about their physical behavior, inferentially justified in believing propositions about the future based on what we know about past correlations of properties, and inferentially justified in believing propositions about theoretical entities in science based on what we know about the behavior of macrobodies, the pivotal question on which the skepticisms stand or fall becomes straightforward. To refute these skepticisms the inferential internalist must claim that the inferences in question are legitimate, and to be legitimate we must be aware of at least a probabilistic connection in each of the cases between the propositions that constitute the available evidence and the propositions inferred from that evidence.

If you recall our recursive analysis of inferential justification, for the above inferences to be legitimate it is *not* necessary that we have a noninferentially justified belief that propositions describing certain sensations make probable certain propositions about the physical world, that propositions describing certain present conscious states make probable propositions about the past, that propositions describing physical behavior make probable the occurrence of certain mental states, that propositions about past correlations make probable propositions about future correlations, or that propositions describing macro-objects make probable propositions describing unobservable entities. It is *in principle* possible to *infer* the existence of probabilistic connections from yet another body of evidence. Perhaps I can construct a nondeductive argument for the conclusion that the occurrence of certain sensations makes probable the existence of the table, for example.

In chapter 4 we made a distinction between primary and secondary epistemic principles. An epistemic principle is a proposition asserting a probabilistic connection between propositions. Within a traditional internalist version of foundationalism, one can define secondary epistemic principles as those that can be justified only inferentially. So we might be able to infer from the fact that litmus paper turned red in a solution that the solution is acidic, but the proposition asserting a connection between its being true that the litmus paper turned red and its being true that the solution is acidic is a *contingent* proposition of a sort that would require inferential justification. The most obvious way to establish a connection between these phenomena is to use some sort of *inductive* argument. The principle that when litmus paper turns red in a solution the solution is acidic is at best a *secondary* epistemic principle.

The internalist foundationalist should understand a *primary* epistemic principle as one that can be justified noninferentially. But are there any uncontroversial examples of primary epistemic principles understood this way? For the inferential internalist this question is absolutely fundamental. According to the inferential internalist, unless we can be noninferentially justified in believing at least some propositions of the form '*E* makes probable *P,*' there will be no inferentially justified beliefs. Furthermore, the inferential internalist's ability to refute skepticism is clearly going to turn on the possibility of discovering a *variety* of plausible candidates for primary epistemic principles. Again, one does not necessarily need a different epistemic principle to bridge each of the problematic "gaps" that give rise to the skeptical challenge. One can always hope, for example, that if one solves the epistemic problems involving perception, the past, and inductive inference, one can employ the relevant information provided by such reasoning to solve the problem of other minds. It is not *completely* obvious, for example, that one cannot develop an inductive justification for believing that the physical behavior of others indicates the presence of conscious states. After all, if one solves these other epistemic problems, one can in one's own case establish a connection between one's own behavior and mental states.[3] Still, if the skeptic has convinced us of anything, it seems clear to me that one will not be able to inductively justify the reliability of inductive inference. And although people keep trying, the idea of constructing a deductive argument in support of the reliability of inductive inference seems prima facie implausible. As we shall see, reasoning to the best explanation is viewed by many as a powerful source of rational belief, and some would attempt to construe inductive reasoning as implicitly relying on reasoning to the best explanation. We shall discuss reasoning to the best explanation in some detail shortly, but for now let us observe that a principle of induction is one of the most obvious candidates for a primary epistemic principle.

Without justified beliefs about the past, however, one cannot even use inductive reasoning, for one will have no premises from which to inductively infer any conclusions. And again the skeptic has surely established that *if* inferential internalism is presupposed, one cannot inductively establish the reliability of memory. So, minimally, I suspect that the inferential internalist will need a noninferentially justified belief in some epistemic memory principle to go with noninferentially justified belief in an inductive principle. Whether or not one needs additional epistemic principles to resolve skeptical problems concerning belief about the physical world is a much more complicated matter, because it depends very much on how one understands propositions asserting the existence of physical objects. Radical phenomenalism, which defines physical objects in terms of facts about what sensations a subject would have were that subject to have other sensations, opens the door to inferentially justified beliefs about the physical world that employ only epistemic principles of memory and induction. It is perhaps an understatement to suggest that radical phenomenalism is not the most popular view concerning the analysis of physical objects, and the more the content of physical propositions goes beyond the world of sensation, the more likely we will need additional epistemic principles to refute the skeptical challenge. The case of theoretical entities is much like the problem of perception. If one can make good on reduction programs, one will need fewer epistemic principles than if the content of theoretical hypotheses goes well beyond even hypothetical propositions about observables.

But before we examine any of these questions more closely, we should return to the fundamental question. Can one plausibly claim to know or justifiably believe *without inference* any propositions of the form '*E* makes probable *P*'? Well, if entailment is simply the upper limit of making probable, then entailment is, of course, one plausible candidate for a relation holding between propositions that one can know directly. But virtually all foundationalists (internalists as well as externalists) agree that one is not going very far beyond the foundations of knowledge and justified belief employing only deductive inference. In short, the only way for the inferential internalist to avoid *massive* (not necessarily global) skepticism is to find a relation weaker than entailment that holds between our foundations and the propositions we infer from them, a relation that we could discover noninferentially. But what would such a relation be? The answer depends critically on how one understands the concept of probability relevant to epistemology and, in particular, relevant to the understanding of epistemic principles.

Inferential Internalism and Frequency Conceptions of Epistemic Probability

By far the most familiar concept of probability is that defined in terms of frequency. On the crudest version of the theory, we say that *a*'s being *G* is probable relative to its being *F* when most *F*'s are *G*. Just as we saw that reliability cannot plausibly be defined in terms of actual frequencies, so too, a slightly more sophisticated frequency theory of probability will no doubt turn to counterfactuals. The problems facing a frequency theory of probability are very similar to the problems facing the classical regularity theory of law. According to the crudest regularity theory of law, it is a law of nature that all *F*'s are *G*'s if and only if there is no *F* that is not *G*. But as we remarked in chapter 4, virtually everyone agrees that some regularities are not *lawful* regularities. It is true that all mermaids have pink eyes, that all of the people in this room are under seven feet tall, that all philosophers with a scar over their right eye caused by a hockey puck that deflected off the stick of a defenseman who had red hair and a tattoo on his right arm are presently at work on a book about skepticism. But none of these true universal statements are laws of nature. Just as mere regularity is not sufficient for lawful regularity, so too, mere statistical regularities are not sufficient for probabilistic connection, even on a frequency theory of probability. It is tempting to solve both the problem of distinguishing laws of nature from accidental regularities and the problem of distinguishing statistical facts that are relevant to probabilistic claims from those that are not, using subjunctive conditionals. It is a law of nature that all *F*'s are *G*'s only if it is true that if anything were an *F* it would be a *G*. It is true that *a*'s being *G* is probable relative to its being *F* only if, were a great many things *F*, they would for the most part be *G*. The relation asserted by contingent subjunctive conditionals is hardly transparent, and I would not dream of suggesting that one can provide an adequate philosophical solution to these problems using subjunctives. If anything, I think the analysis of subjunctive conditionals will be, in the end, parasitic upon a plausible analysis of the distinction between laws of nature (universal and probabilistic) and accidental regularities.[4] But for the moment let us suppose that we have an adequate understanding of the sort of statistical regularities that are relevant to probability given a frequency interpretation of the concept.

Although the details of the account are far from unproblematic, the basic idea of a frequency theory seems clear. It is less obvious how one would incorporate frequency conceptions of probability into a discussion of epistemic principles. How can we understand in terms of frequency the claim that a proposition *E* makes probable a proposition *P*? One thing is clear and that is that we must introduce into the account of *E*'s making probable *P* a *class* of propositions or proposition pairs. If we try to capture the sense of probability relevant to epistemic principles (the sense of probability involved in the principle of inferential justification) using the concept of frequency, I suppose we might suggest that in claiming that *P* is probable relative to *E* we are simply asserting that *E* and *P* constitute a pair of propositions, which pair is a

member of a certain class of proposition pairs such that, when the first
member of the pair is true, usually the other is. Thus, in claiming that
my seeming to remember eating this morning (*E*) makes it likely that I
did eat this morning (*P*), I could be construed as asserting that the pair
of propositions *E/P* is of the form '*S* seems to remember *X/X*,' such that
most often when the first member of the pair is true, the second is. For
simplicity, I explain the view returning to the crudest conception of
frequency defined in terms of mere statistical regularity. Again, more
plausible versions of the view will almost certainly employ subjunctives. For my seeming to remember *X* to make it probable that *X*, it must
be true not that most memory experiences are veridical but that if one
were to continue to have indefinitely many memory experiences, most
of them would be veridical.

If we adopt something like this frequency conception of *epistemic*
probability, then one of the most important tasks of normative epistemology would be the empirical task of discovering which epistemic
principles are true. Ideas in philosophy are seldom completely original,
and it is interesting to note that the current metaepistemological debate
over reliabilism very closely parallels a discussion of epistemic probability by Russell over forty years ago in *Human Knowledge: Its Scope
and Limits*. In one of the most illuminating philosophical examinations
of probability ever undertaken, Russell insists that for our conclusions
to be rational our evidence must make them probable in a frequency
sense of probable.[5] This conclusion involved a considerable evolution
in his thought, because the earlier Russell implicitly adopted a radically
different account of probability. The later Russell was motivated by
many of the same concerns that move reliabilists. Like the reliabilist,
Russell was convinced that for epistemic concepts to be of philosophical interest they must be conceptually tied to truth. And he was further
convinced that only a frequency conception of probability could secure
a conceptual connection to truth. Both Russell's crucial concept of
probability and the reliabilists' crucial concept of reliability are defined
in terms of frequencies (at least initially). Indeed, it may not be too far
a stretch to suggest that the Russell of 1948 was implicitly one of the
first externalists, because he must have realized that once one defines
epistemic probability in terms of frequency, there will be no nondeductive primary epistemic principles (that is, nondeductive epistemic principles known noninferentially). But if there are no nondeductive primary epistemic principles and one wants to remain a foundationalist,
one must either give up inferential internalism or embrace fairly radical
skepticism.

This last point is so important that it deserves emphasis. If one defines epistemic probability in terms of frequencies, then the inferential internalist faces a virtually insurmountable problem in the attempt to stave off local skepticisms. Nondeductive epistemic principles on a frequency interpretation of epistemic probability are certain to be very complex *contingent* truths, and even the most daring foundationalist will be unwilling to claim direct or immediate awareness of the frequencies that must obtain in order to make them true. But if the foundationalist must *infer* all epistemic principles from evidence—that is, if all epistemic principles are secondary epistemic principles—then there will be no escape from the specter of the vicious regress raised by the second clause of the principle of inferential justification.

Epistemic Principles as Analytic Truths

If one accepts the conclusion that the inferential internalist will be inevitably led to skepticism by understanding *epistemic* probability in terms of frequencies, an understanding that renders epistemic principles *contingent*, one might consider the possibility of construing epistemic principles as *analytic*. Propositions describing sensations make probable, or make it rational to believe, propositions about the physical world because that is how we *understand* or *define* rational belief. Propositions describing memory as evidence for the existence of the past, past correlations as evidence for the future, and the behavior of bodies as evidence for mental states are all analytic, are all really implicit assertions about the *concept* of rationality.[6] We can have noninferential knowledge of epistemic principles because we can have noninferential knowledge of analytic truths. On an acquaintance theory, the source of such knowledge might be direct acquaintance with our thoughts and the relations that hold between them.

Pollock defends a closely related attempt to solve epistemological problems through "conceptual analysis."[7] He argues, in effect, that rational belief is to be understood in terms of belief to which one is entitled. He goes on to claim that one should include in the very meaning-rules governing the use of concepts normative claims about the circumstances in which one is permitted to assert propositions employing the concept. The very concept of a physical object just is the concept of a thing whose existence we are entitled to assert as a result of having certain sensations. The very concept of someone's being in pain includes reference to the propriety of concluding that someone is in pain on the basis of exhibiting pain behavior. Pollock's idea here bears at least a family resemblance to the later Wittgenstein's pervasive concept of the nondefining (but noncontingent) criterion.

There are many difficulties with attempts to make epistemic principles analytic, but here I will stress only two. First, it seems to me a little difficult to suppose that the many skeptics and those who took them seriously were all simply misusing language. However implausible we might view skepticism about the physical world, are we really to maintain that such skeptics were simply contradicting themselves? Can we really dismiss the skeptical challenge by exclaiming that we just *understand* rationality in such a way that it follows from the concept alone that sensations make it rational to believe propositions about the physical world? In short, the solution seems too easy. It seems that one cannot accommodate the undeniable force of the skeptical challenge within the framework of this attempt to find epistemically reasonable epistemic principles. This complaint will no doubt fall on at least some deaf ears, particularly when addressed to those philosophers who never have felt the force of the skeptical challenge and who have instead always viewed skepticism as a view gone so far wrong that it does not deserve to be taken seriously. To them I would stress the second objection.

One is sorely tempted to suppose that philosophers who take epistemic principles to be analytic do normative epistemology by simply listing their prephilosophical beliefs, deciding what they do infer the propositions believed from, and proclaiming the epistemic principle sanctioning such inferences to be analytic. But what exactly do all of these inferences have *in common* that makes it plausible to claim that they fall under a single concept of rational inference? Is the concept of probabilistic or evidential connection simply a disjunctive concept? It is rational to believe *P* on the basis of *E means*: EITHER *P* is a proposition about the physical world inferred from truths about sensation OR *P* is a proposition about the past inferred from memory states OR *P* is a proposition about the future correlation of properties inferred from past correlations of properties OR (for theists only) *P* is a proposition asserting the existence of God inferred from a feeling of wonderment at the complexity and beauty of nature OR (for palm readers only) *P* is a proposition about one's future life inferred from the nature of the lines on one's palm OR. . . . Of course it is possible to artificially define a disjunctive concept, but it seems to me absurd on the face of it to suppose that so pervasive and important a concept as making probable or making rational is an ad hoc disjunction defined in terms of pairs of proposition types. It surely makes sense to ask: "In virtue of what do both sensations and memory experiences make probable, respectively, propositions about the physical world and propositions about the past?'' The frequency theorist at least has an answer: it is in virtue of a high frequency with which there is truth preservation—when certain propositions of one sort are true, usually certain propositions of the other sort are true. But as we have seen, the frequency theorist makes these propositions asserting truth preservation contingent, and contingent in a way that seems to eliminate the possibility of ending the crucial regress within the framework of inferential internalism.

Epistemic Principles and Subjective Conceptions of Probability

The most convenient concept of probability to employ in epistemological refutations of skepticism is a so-called subjective conception of probability.[8] Crudely, the subjectivist will identify the probability P has for S with the degree of confidence S has that P is true. The more confidence S has in P, the higher the probability of P for S. Given this approach, the probability will be relativized to an individual. You and I might possess the same evidence for P while the probability of P for you is different from the probability of P for me.

Epistemic principles, of course, relate propositions. There are at least two ways in which a subjectivist might try to understand propositions of the form 'E makes P probable for S.' One sort of subjectivist will understand the degree to which E makes probable P in terms of the way in which belief in E affects belief in P. So on this view, E will make P probable for S when S's belief in E causes S to believe P (ceteris paribus). The more strongly S believes P when S believes E, the stronger will be the probabilistic connection between E and P for S. The other sort of subjectivist might understand epistemic principles in terms of beliefs about probability. Roughly, the idea is that E will make P probable for S if S believes that E makes P probable. Now such an account of epistemic principles would appear to be pathetically circular. One invokes the concept of probability in specifying the object of the belief that makes true the relativized principle of probability. But the subjectivist can circumvent this objection by insisting that the probability which is the subject matter of the belief is our old frequency conception of probability. The *epistemic* probability of P is high for S relative to E when S believes that in a frequency sense E makes P probable.

These subjectivist accounts of probability are exceptionally crude. They can be made more sophisticated in a variety of ways. Although he expresses his view primarily in terms of epistemic rationality, Richard Foley, for example, is *essentially* a subjectivist with respect to epistemic principles, but he insists that the attitudes that define epistemic rationality are those that a subject has after a process of ''ideal'' reflection, and he goes to great length describing the conditions that constitute ideal reflection.[9] Bayesians typically start out with a subjectivist conception of probability (the prior probabilities they plug into their formula are usually understood subjectively) but argue that there are constraints on how the rational person combines subjective probabilities. The constraints are imposed by principles of probability such as Bayes' theorem.

Earlier we discussed in detail the position I called epistemic commonsensism. To discuss skepticism within the framework of a subjectivist conception of probability is essentially to beg the question. The skeptic knows that people believe all of the propositions under skeptical attack. In calling into question the rationality of these beliefs, the skeptic is hardly calling into question the fact that they are held. The question is whether we possess justification that makes probable the truth of what we believe. And this question is not to be answered by determining whether or not we do in fact believe one set of propositions as a result of accepting others, or believe that in a frequency sense one set of propositions makes probable another.

The fundamental objection to subjectivist conceptions of probability is similar to the fundamental objection to viewing epistemic principles as analytic. We are surely not contradicting ourselves when we decry as irrational, people who believe *P* on the basis of *E*. We may recognize that it is a deep-rooted fact about human psychology, for example, that when people fear death they believe much more strongly in an afterlife. But the skeptic is surely not contradicting himself if he argues that despite the psychological facts, belief in an afterlife based on fear of death is irrational. It is an open question as to whether the alleged evidence here makes probable the conclusion. The causal connection between beliefs does not force us on pain of contradiction to recognize a probabilistic connection of the sort relevant to epistemology. The second sort of subjectivist discussed earlier will argue that fear of death makes epistemically probable for *S* an afterlife only if *S* believes (perhaps after a process of reflection) that fear of death does make probable (in a frequency sense) the existence of an afterlife. But again, can one seriously argue that the skeptic who lives in a society of people odd enough to believe in such a statistical correlation cannot without contradiction deny that this conclusion is probable for these people relative to their evidence?

This argument is not completely fair, perhaps. I know that philosophers like Foley emphasize the importance of different perspectives. When I, the skeptic, criticize "your" epistemic principles, perhaps I am implicitly making claims only about what epistemic principles hold for me. But even after we make all of these distinctions, it is difficult to rid oneself of the feeling that the subjectivist has consigned epistemology to psychology. If we do understand epistemic probabilities subjectively, the philosophical enterprise of epistemology has been at the very least misconceived. I am certain that the classic figures who worried about the rationality of their beliefs were not merely interested in finding out the deep-rooted psychological facts about what they were disposed to believe under certain conditions. It was the rationality of these very dispositions that was the object of their concern.

It is also worth pointing out, perhaps, that sophisticated subjectivist conceptions of epistemic principles might not in the end help very much with the refutation of skepticism. If what makes it true that E makes P probable for me is some complicated fact about what I would believe under certain circumstances *and* inferential internalism is true, then for me to be justified in believing P on the basis of E I would still have to be justified in believing that E makes P probable (for me). Justified belief in this proposition requires justified belief in a complex contingent counterfactual. It is unlikely that the internalist will allow noninferentially justified belief in the counterfactual. It will need to be inferred from some other proposition F. But now I must be justified in believing that F makes probable the counterfactual that defines E making probable P. Whether or not F does make probable that E makes probable P (for me) will be determined by another complex counterfactual about what I would believe, and so on ad infinitum. Because a Foley-style subjectivist will probably have no noninferentially justified beliefs in epistemic principles, we may seem to be involved in a vicious regress. One might reply that the regress need not be regarded as vicious, for the conditions that define for me the rationality of the (infinitely many) epistemic principles are *all* explicated using counterfactuals. But the antecedents of these counterfactuals refer to what I would think about ever increasingly complex propositions, and it seems to me that a finite mind will be incapable of even considering the content of such propositions beyond a certain level of complexity.

Epistemic Principles as Synthetic Necessary Truths and the Keynesian Concept of Epistemic Probability

Is there any other way of avoiding relatively massive skepticism for the inferential internalist? It seems to me that the answer is yes *only* if we can understand the concept of nondeductive epistemic probability as being much more like the concept of entailment, and can subsequently convince ourselves that epistemic principles are necessary truths knowable *a priori*. And if we reject this attempt to make epistemic principles analytic, the only way to construe them as necessary truths knowable *a priori* is to accept a Keynesian concept of making probable as an *internal* relation holding between propositions.[10] Let me be more explicit.

We may define an *internal* relation as one that holds necessarily because of the nonrelational character of the relata. If X stands in an internal relation R to Y, then the intrinsic (nonrelational) character of X and Y is sufficient for the relation R to obtain. One might argue, for example, that 'being darker than' is an internal relation that holds between the colors black and white. It is a necessary truth that if black and white both exist, then black is darker than white. The relation 'darker than' holds between black and white solely by virtue of the intrinsic character of the respective colors. 'Being a lower note than' is arguably an internal relation that holds between middle C and middle E on the piano.

If there are internal relations and we can be directly acquainted with the intrinsic character of the relata of such relations, we might also be directly acquainted with the fact that the relation obtains. If propositions are the sorts of things we can hold directly before our minds, and if making probable is an internal relation holding between propositions, it might not be that hard, dialectically, to claim that one can hold directly before one's mind the kind of fact that makes propositions of the form 'E makes probable P' true. Epistemic principles will become synthetic necessary truths knowable *a priori*. The relation of making probable will be very much like the relationship of entailing. Both relations are internal relations holding between propositions and both are knowable *a priori*.

Of course, even if one accepts such a view, one will certainly want to stress the differences between a logical relation of making probable and a logical relation of entailment. The logic of epistemic probability is different from the logic of entailment. From the fact that P entails Q it follows that P conjoined with any other proposition entails Q. From the fact that P makes probable Q it does *not* follow that P conjoined with any other proposition makes it probable that Q. Entailment is transitive. If P entails Q and Q entails R, then P entails R. The relation of making probable is not transitive. If P makes probable Q and Q makes probable R, it does not follow that P makes probable R. There *appear* to be interesting analogies between deductive and probabilistic reasoning, but there are a number of technical difficulties that usually arise. It seems plausible initially, for example, to argue that there are nondeductive analogues of both modus ponens and modus tollens. The principle that if *P* makes probable *Q* and *not-Q* then (ceteris paribus) probably *not-P* would seem to be prima facie plausible. But from the fact that if I am entered in a fair lottery I will probably lose and I do not lose, it is difficult to conclude *on that basis* that the lottery was probably not fair. Despite disanalogies between entailment and a Keynesian concept of making probable, however, the crucial similarity remains. Both are internal relations holding between propositions and both relations can be known to obtain *a priori*.

I have said almost nothing until now about how a foundationalist will understand foundational knowledge of necessary truth. I take it to be one of the great virtues of an acquaintance theory that it can provide a *unified* account of noninferentially justified belief and knowledge. According to many philosophers, there is an enormous difference between *a priori* and *a posteriori* knowledge. BonJour, as we saw, was driven to a *radically* different account of *a priori* and *a posteriori* knowledge. But on an acquaintance theory there is essentially no difference between the *source* of noninferential empirical knowledge and noninferential knowledge of necessary truth. In both cases the source of knowledge is acquaintance. The only real difference is the nature of that with which one is acquainted. When I know that black is darker than white or that bachelors are unmarried, the source of my knowledge is direct acquaintance with relations that hold between either thoughts or properties (depending on whether you take necessary truths to hold by virtue of relations that obtain between thoughts or properties). When I know noninferentially that I am in pain, I am directly acquainted with my being in pain. There is a philosophically important distinction to be drawn with respect to the objects of *a priori* and empirical knowledge. A relation of being darker than holding between black and white and a relation of "containment" holding between being a bachelor and being unmarried, are facts importantly different from the fact that I am in pain. The former facts exist in all those worlds in which black and white, and being a bachelor and being unmarried, exist. The latter fact does not obtain in all of those worlds in which both I and the property of being in pain exist. But for all the importance of distinguishing two kinds of truths, an acquaintance theory can accomplish the important goal of making clear what is *common* to the two kinds of knowledge that makes it appropriate to regard both as noninferential.

I will put my tentative conclusion as starkly as I can. If you are an inferential internalist, that is, you accept the second clause of the principle of inferential justification, then you must hold that in the sense relevant to epistemology, making probable is an internal relation holding between propositions, and that one can be directly and immediately acquainted with facts of the form '*E* makes probable *P*.' Otherwise, you must embrace massive skepticism with respect to the past, the external world, the future, and other minds. But are we directly acquainted with an internal relation of making probable holding between propositions? Is the concept even intelligible, and if it is, is it plausible to understand *epistemic* probability in terms of such a relation?

In examining this question it is useful to turn again to Russell's (1948) discussion of epistemic probability—he calls it credibility. As I indicated earlier, Russell is a particularly interesting case because later in his career he attacks the very Keynesian concept of probability he once implicitly endorsed. Although most philosophers might remember Russell's famous discussion of the problem of induction (Russell 1959) most for his statement of the problem, it is important to realize that he also attempted to solve that problem. Toward the end of the chapter devoted to induction, Russell introduces what amount to epistemic principles of induction and claims that one can know these principles to be true without inferring them from any evidence. One of the principles he states this way: The more often *A* and *B* have been found associated with no failures of association the more likely it is that when we have a fresh case of *A* we will have a case of *B* (where *A* and *B* are properties). Of this principle Russell says that experience can neither prove nor disprove it.[11] There is no *evidence* one can put forth in support of a principle of induction. But we can know it nevertheless. We can regard it as one of those innumerable "first" truths upon which the rest of our reasoning depends.[12] Although he was not as clear as he might have been, it seems obvious that he was adopting the position that the principle of induction is a synthetic necessary truth knowable *a priori*. But he could only plausibly maintain such a position, as we have seen, if he at least implicitly adopted a Keynesian conception of epistemic probability.

Why did Russell abandon the Keynesian view? As I indicated earlier, the answer lies with his later insistence that the epistemic concept of making probable be closely connected to the concept of *truth*. Again, it is difficult to overstate the similarities here between the later Russell and contemporary externalists. Philosophers would only *care* about the concept of epistemic rationality, Russell maintains, if epistemic rationality were in some way tied to truth. The ultimate goal of the philosopher is to have true beliefs, and one only concerns oneself with having rational beliefs insofar as one is entitled to believe that there is a connection between having rational beliefs and having (mostly) true beliefs. But the fundamental problem with a Keynesian conception of probability is that one loses that necessary connection between epistemic rationality and truth. Suppose, for example, that a Keynesian makes the claim that the proposition that I seem to remember X makes probable the proposition that X occurred. This relation of making probable is a sui generis internal relation holding between propositions. It can obtain if it is *never* the case that when one seems to remember X, X occurred. But if there is no necessary connection between E making P probable and propositions like P usually being true when propositions like E are true, why should I bother to conform my beliefs to what the evidence makes probable? At the very least (as BonJour would argue) we would need some independent reason to think that when E makes P probable in the Keynesian sense, *usually* the relevant frequencies obtain. But if we accept this suggestion we are off and running on the very regress of justification that introduction of a Keynesian conception of probability is designed to contain.

The above argument is a double-edged sword. The very defect Russell pointed to will be viewed by at least some internalists as one of its virtues. Recall that most internalists shared a strong intuition that in the demon world our beliefs about physical objects are perfectly rational even if they are uniformly false. Or consider Russell's famous skeptical scenario about the past. Russell argued that it is perfectly imaginable that we all came into existence a few minutes ago replete with detailed and vivid memories of a past life.[13] Let us also suppose that we cease to exist a few minutes later. In this possible world, the vast majority of beliefs about the past based on memory are false. Memory experiences, on a frequency conception of probability, will not make probable for us any truths about the past. But there is an equally strong intuition shared by many philosophers that in this possible world beliefs about the past would be perfectly rational. They would be just as rational as *our* beliefs about the past, for the evidence supporting their beliefs would be identical to the evidence at our disposal. We seem to be on the brink of returning to the now familiar stalemate involving internalists and externalists concerning fundamental desiderata for a plausible analysis of the concepts fundamental to epistemology.

Can the Keynesian accommodate Russell's concern that epistemic probability be relevant to truth seekers? I worry that in the end the answer might still be no. But the Keynesian can certainly engage in a conversation that does at least obscure the issue. Imagine the following dialogue:

Russell: So this odd relation of making probable doesn't have any necessary connection to truth. Why then should the epistemologist be concerned with it?

Keynes: Of course it has a necessary connection to truth. When *E* makes probable *P* and *E* is the case, then *P* is *probably* true. No one wants it to turn out that it is a necessary truth that when *E* makes probable *P* and *E*, then *P*. The most we can ask for is a probabilistic connection. I don't understand your concern.

Russell: But on your view my seeming to remember *X* can make it probable that *X* even if it is *never* the case that memory is veridical.

Keynes: Sure, but if your seeming to remember *X* makes it probable that *X*, then it is of course *probably* the case that memory is usually veridical and *probably not* the case that it is always illusory.

Russell: But what reason do you have for thinking that memory is usually veridical?

Keynes: There is a sui generis internal relation of making probable holding between the proposition that I seem to remember *X* and *X*—my remembering *X* makes it likely that *X*.

Russell: But what reason do you have for thinking that when that relation holds, memory is usually reliable?

Keynes: My awareness of that relation holding *is* my reason for thinking that memory is usually reliable. If the relation obtains, then the occurrence of memory makes likely for me its veridicality. You want more from probability than probability can give. You were *right* in *The Problems of Philosophy* when you argued that experience can neither confirm nor disconfirm the principle of induction. And you were right in insisting that, for all that, inductively justified beliefs about the future are perfectly rational because, however the world turns out, it is still epistemically probable that past correlations continue into the future.

When the rhetorical dust settles, it seems clear that a Keynesian analysis of epistemic probability has the following implications. First, propositions asserting probabilistic connections are necessary truths, knowable *a priori*. If one can hold before one the propositions probabilistically connected, one can be directly acquainted with the probabilistic connection. Second, from the fact that *E* makes probable *P* nothing follows about the actual frequency with which *P*-type propositions are true when *E*-type propositions are true. Because we are using epistemic probability in the analysis of inferential justification, it follows that there is no necessary connection between having inferentially justified beliefs and having mostly *true* beliefs. It is not only true (as it should be) that a single inferentially justified belief might be false (when the inference is nondeductive), it is also true that *all* inferentially justified beliefs might be false (when the inferences are nondeductive).

Lest this last consequence seem to you too worrisome, let me remind you that *sophisticated* externalist metaepistemologies have precisely this same result. A reliabilism that does not fall prey to obvious objections cannot understand reliability in terms of *actual* frequencies. The reliabilist will almost certainly turn to the concept of a propensity or employ subjunctive conditionals describing what would be the outcome of indefinitely long use of a belief-producing process. On either view, it will no longer be necessarily true that the majority of justified beliefs are true. It will be epistemically *probable*, according to such reliability theories, that the majority of justified beliefs are true, but as we have just seen, the Keynesian can insist on this tautological claim as well.

There are only two metaepistemologies that are going to secure a strong connection between justification and truth. The first is the Cartesian conception of justification that precludes possibility of error. The second is a *crude* externalism that defines justification in terms of actual statistical correlations between having a belief of a certain sort and the belief's being true. No matter how we interpret the modal operator contained in the Cartesian conception of justification, the view will imply that the range of justified beliefs is highly restrictive. There will be no difficulty establishing relatively massive local skepticisms within such a framework. Externalist analyses that rely on crude statistical frequencies will be vulnerable to devastating counterexamples.[14]

The Dialectical Alternatives

At this point the choices should seem relatively clear. It seems to me that the epistemologist who wishes to avoid massive skepticism must choose between inferential externalism and a Keynesian conception of probability. If the choice is inferential externalism, then you should regard your defense of the metaepistemological position as your last epistemological task qua philosopher. If you embrace the Keynesian conception of probability, you need to figure out what probabilistic connections you are directly aware of. And as part of that task, you must of course convince yourself that you really are acquainted with this relation of making probable holding between propositions.

The Search for A Priori Epistemic Principles

The foundationalist who embraces both clauses of the principle of inferential justification and seeks to avoid both epistemic and conceptual regress concerning justified beliefs about probabilistic connections by embracing a Keynesian conception of epistemic probability, will refuse to offer an argument for the various epistemic principles (principles asserting probabilistic connections) he endorses. It is, after all, part of the view that one can discover a priori *primary* epistemic principles. One worries, at this point, that normative epistemology will reduce to a "take it or leave it" catalogue of such principles advanced by the Keynesian foundationalists. And since almost all epistemologists are determined to avoid skepticism, one can be sure that the epistemic principles "discovered" are just those that will secure justification for at least most of those "commonsense" beliefs we held prior to our epistemological investigation. It does not follow, however, that because a philosopher is committed to the a priori status of a principle one would beg the question against such a philosopher by arguing *against* the principle. One can, after all, discover that a proposition one thought was necessary and known a priori is false. And one way to make such a discovery is to realize that a principle has certain obviously unacceptable consequences.[15] This book is concerned primarily with metaepistemological questions and the implications their answers have for the approach one should take to the traditional skeptical challenge, and so even if I thought the Keynesian conception of a probability relation were correct, I would not attempt a detailed list of epistemic principles. Furthermore, as my concluding comments will indicate, I think that there is a distinct possibility that the probability relation philosophers *seek* is a chimera. Still, at this point it might be useful to discuss briefly the kind of dialectical argument that one will engage in if one embarks on the search for Keynesian probability relations.

Principles of Formal Probability

What are the most obvious candidates for necessary a priori principles asserting probabilistic connections? An obvious place to begin is the well-known probability calculus already in place. We all learned, for example, how to determine the probability of conjunctions and disjunctions once we know the probability of their atomic propositions (and their probabilities relative to one another). Why not simply list these among the principles of probability that the Keynesian will employ? The answer to this question is not straightforward. For one thing, it is likely that the principles of probability advanced by formal probability theory deal with something much closer to a *frequency* conception of probability. Indeed, the proof that the probability of a conjunction (with probabilistically unrelated conjuncts) is the product of the probabilities of the conjuncts proceeds straightforwardly only on a frequency conception of probability. It may be that for every necessary truth concerning probability on a frequency interpretation, there will be a corresponding necessary truth asserting a Keynesian probability relation—it may be, for example, that the proposition that P and Q each have a .5 probability makes it likely (in the Keynesian sense) that the conjunction (P and Q) is false.

Whether or not all of the necessary truths concerning probability understood in terms of frequency have Keynesian probability assertions as corollaries, it seems clear that such principles will not take us very far in resolving traditional problems of skepticism. And the obvious explanation is just that one needs to know the probabilities of atomic propositions before one can determine the probabilities of truth-functionally complex propositions made up out of these atomic propositions. The foundationalist's candidate for a proposition whose probability can be determined without knowing the probability of another proposition is, of course, a proposition known directly through acquaintance. But as the skeptical arguments were supposed to show, the propositions we need to justifiably believe in order to avoid rather massive skepticism are *not* truth-functionally complex propositions made up out of propositions known directly. Even the most primitive forms of phenomenalism recognize the need to introduce non-truth-functional counterfactuals into any plausible analysis of propositions about the past, the physical world, and other minds. In short, if we are to use a Keynesian conception of probability to avoid skepticism, we cannot restrict ourselves to epistemic principles that one could prove a priori even when the probability is given a frequency interpretation.

Furthermore, it seems reasonable to focus our initial search on principles whose application does not presuppose the availability of other principles. This approach will push into the background two of the "magic bullets" often advanced by epistemologists as the cure-all for epistemic problems—Bayesian reasoning and reasoning to the best explanation.

Bayesian Reasoning

Bayes' theorem gives one a way of determining the probability of a hypothesis H relative to evidence E when one already knows the *prior* probability of *H* (presumably relative to some background evidence), the prior probability of *E* given *H*, and the prior probability of *E* (relative to background evidence). One determines the probability of *H* given *E* by multiplying the prior probability of *H* by the probability of *E* given *H* and dividing it by the prior probability of *E*. Most Bayesians understand the prior probabilities subjectively, but if we are serious about epistemology and reject the basic idea behind a coherence theory of justification, we will not be content to use a principle to make relevant adjustments to the degree of our commitment to various propositions. We will need some sort of account of what gives the relevant hypotheses their prior probabilities. And this will require some way of determining the probability of the hypothesis *H* relative to our background evidence. It will also require some way of determining the probability of the evidence *E* being available given the hypothesis, and unless *H* entails *E* this will no doubt require yet another application of some nondeductive principle of reasoning.

It should also be noted in passing that there are formal difficulties with Bayesian reasoning. Consider the case of a lottery, for example. Suppose that the prior probability for me of a lottery's being fair is high, .9, say. The probability of my winning the lottery given that it is fair is .001, let us further suppose. If I win, it would seem that I can reach the conclusion that in all probability the lottery was not fair. But this is strongly counterintuitive. It is not clear that this technical problem is insurmountable. We can also reason that all fair lotteries will be vulnerable to a Bayesian refutation of their fairness, and once we realize that we cannot use Bayes' theorem to discriminate between fair and unfair lotteries, we might be able to use the theorem itself to reach the conclusion that one should not use the theorem in such contexts (much as one can rely on memory to distrust memory under certain circumstances).

Nevertheless, the fundamental problem with Bayes' theorem as a tool for resolving our classical skeptical problems is that the skeptic will argue that the relevant probabilities one needs to "plug into" the formula are unavailable. I want to know whether the table is before me when I have this kind of sensation. What is the prior probability of there being a table before me? What is the prior probability of my having a sensation of this kind relative to there being a table before me? And what is the prior probability of my having a sensation like this? Discovering these probabilities would be no easier than discovering the conditional probability of the table's existence on the occurrence of this kind of sensation, the very problem with which we began.

Reasoning to the Best Explanation

Reasoning to the best explanation is offered by some as another all-powerful weapon for use against the skeptic.[16] Thus Locke, and sometimes Russell, seemed to argue that we are justified in positing physical objects as the best explanation for the order and connections that we find in experience.[17] Past events can be introduced as the best explanation for the occurrence of present memory experiences. We are justified in believing that others have minds precisely because positing their mental states would best explain their observed behavior.

To assess these proposed solutions to our skeptical problems we need, of course, a detailed conception of reasoning to the best explanation. It is worth emphasizing, however, that historically, reasoning to the best explanation seems to be put forth as a solution to skeptical challenges from *within* an inferential internalist's framework. As we have seen, if one embraces inferential externalism it is difficult to see why one would have need of anything as abstract as reasoning to the best explanation. The externalist has no motive to try to subsume various fundamental belief-forming processes under some more general pattern of argument like inductive reasoning or reasoning to the best explanation.

I have discussed elsewhere in some detail the limitations of reasoning to the best explanation as a way of answering the skeptical challenge.[18] Therefore I will sketch here only the problems reasoning to the best explanation faces. As I said earlier, we need a model of reasoning to the best explanation. If we are Keynesians, we need an argument form whose premises stand in the relation of making probable to its conclusion. I have argued that the most sophisticated form of reasoning to the best explanation will look something like the following:

1. O (a description of some observed phenomena)
2. Of the available and competing explanations $E1$, $E2$, ..., En capable of explaining (if they were true) O, $E1$ is the best according to the correct criteria $C1$, $C2$, ...,Cn for evaluating explanations.
Therefore,
3. $E1$.

I presuppose in the following comments that we have a roughly adequate account of the formal requirements for a correct explanation (that is, the requirements excluding the truth of the propositions that constitute the explanans).[19] On virtually any account of what constitutes an adequate explanation, there will be literally infinitely many *potential* explanations for a phenomenon which, if true, would be adequate explanations. The potential explanations for our sensations seem to range from physical objects of the sort posited by common sense to Berkeley's God and Descartes's evil demon.

Now to evaluate the claim that 1 and 2 of our argument to the best explanation make probable 3, we will need to know more about what these criteria for evaluating explanations are. But even before we examine such criteria, we should note that reasoning to the best explanation seems to presuppose that it is likely that phenomena have explanations. Unless we have an antecedently justified belief that most things have explanations, it is difficult to see why we would take the fact that one potential explanation seems better than others as a reason for supposing that the potential explanation is true. And although such a presupposition is perfectly natural, and even appropriate, in the context of scientific investigations (although in these days of quantum theory there seems to be much more tolerance for indeterminacy at the microlevel), in the context of a philosophical refutation of very fundamental sorts of skeptical arguments it is hardly unproblematic. Remember that the skeptic we are concerned with is attacking the possibility of justifiably believing anything about the past based on memory and anything about the external world based on sensation. A Humean skeptic will maintain that it is a contingent empirical fact that any given event has a cause and that it is, of course, contingent that everything, or that most things, have a cause. If in justifying our beliefs about the past or the physical world we must suppose that most events have causes, the skeptic will inquire as to our justification for that supposition. Was it inductively established? If so, how was that inductive justification accomplished without a prior solution to the problem of justifiably believing propositions about the past? Indeed, if we rely on *any* information about the past at all in justifying our belief that most events have causes, we will need an independent solution to the problem of justifying beliefs about the past, and thus if the preceding argument is correct we will have foreclosed the possibility of using reasoning to the best explanation in an attempt to justify beliefs about the past. If in reaching the conclusion that most events have causes we rely on the fact that our sensations have causes, we will again need a prior solution to skepticism about the physical world and we will be precluded from using reasoning to the best explanation in order to get that solution.

Of course, a Keynesian *could* claim that every proposition asserting the occurrence of some event makes probable the truth of a proposition asserting that there is a cause of that event. And this proposition could be taken to be necessary and knowable *a priori*. Like any principle offered by the Keynesian, it is difficult to conclusively refute (more about this later). But it does not seem a very promising candidate for a synthetic necessary truth about probability relations, and in any event, we will be acknowledging the point I am trying to establish, that the use of reasoning to the best explanation will require reliance on more fundamental principles of reasoning.

Another preliminary point worth making is that even if we concede that we can know a priori that everything probably has a causal explanation, there is an important and fundamental ambiguity in the concept of *best* explanation. When we talk about one of a number of competing explanations being the best by certain criteria *C*, we can mean one of at least two different things. First, we may mean only that the explanation in question is better by criteria *C* than each of the competing explanations taken individually. *E1* might be better than *E2*, than *E3*, than *E4*, and so on. *This* fact by itself does not imply that even if we are justified in believing that one of these explanations is correct, we are justified in believing that *E1* is the correct explanation. It is true that if we are forced to "bet," the rational bet would be *E1*. If we are forced to make plans based on some explanation, *E1* is the rational explanation to choose. But being the rational explanation to act on is not the same thing as being the rational explanation to believe. In one sense of rational action we can rationally choose *A* over its alternatives because we rationally believe that *A* has the best chance of maximizing expected utility, meaning only that *A* has a better chance than each of its alternatives taken individually. At the same time, of course, we might cheerfully admit that it is more likely that one of the alternatives will maximize utility than it is that *A* will. Put more precisely, the disjunction whose disjuncts identify each alternative to *A* as being the maximizer is more likely to be true than the proposition that *A* is the maximizer.

Similarly, I can cheerfully admit that *E1* is the most attractive explanation while admitting that the disjunction of propositions asserting alternative explanations is much more likely to be true. In such a situation I stress again that there may be nothing irrational about acting *as if E1* is the correct explanation, particularly if one must do something and one must make plans based on some particular explanation. But as an epistemic agent one certainly does not need to fallaciously infer that *E1* is likely to be the correct explanation.

To refute the skeptic who concedes that we have knowledge that there is an explanation of some phenomenon (memory states or sensations, for example), you must establish that our commonsense hypotheses are more likely to be the correct explanation of these phenomena than the disjunction of all alternative explanations. And it is crucial to recognize how formidable a task this is, particularly given the relative ease with which we can come up with indefinitely many explanations that will satisfy the formal constraints on explanation. Earlier I indicated that the skeptical scenarios may resurface in the context of an attempt to use reasoning to the best explanation to resolve skeptical problems. One of the obvious roles such scenarios play is to generate competing explanations to be included among the indefinitely many disjuncts in that disjunction that has to be less plausible than the hypothesis of common sense.

Keeping all of this firmly in mind, let us examine a few specific proposals of criteria for choosing between alternative explanations. Each suggested criterion can be thought of as a proposed probability principle that might be endorsed by the Keynesian.

Perhaps the single most common virtue of explanation cited by philosophers is simplicity. Other things being equal, the simpler explanation is to be preferred over the more complex. There are, of course, many different ways of interpreting simplicity. Sometimes simplicity seems to be understood in terms of the number of things or events to which the explanation commits one. Sometimes the crucial question concerns the number of *kinds* of things to which the theory commits one. In coming to grips with skepticism about the physical world we can compare our commonsense hypotheses about physical objects to, let us say, a Berkeleyan hypothesis about a very complex mind orchestrating the comings and goings of sensations. Which theory is simpler? Well, Berkeley had just minds, mental states, and causation. The commonsense hypothesis has minds, mental states, causation, and physical objects. On any criteria of simplicity Berkeley seems to win, but despite Berkeley's protestations to the contrary, most of us will conclude that if Berkeley wins, so does skepticism.[20]

Of course, one can introduce other notions of simplicity. Paul Moser introduces the notion of gratuitous elements in an explanation.[21] The details of Moser's account are not completely clear to me, but he seems to want an explanation of sensation, in particular, to accommodate what he calls the *contents* of sensory states. Although I am not sure he would like this way of putting it, it seems to me that his view amounts to the claim that sensations are intentional states that have intentional objects. These intentional objects (which may or may not actually exist) are physical objects. Moser believes, I think, that skeptical scenarios typically leave unanswered "Why?" questions about the intentional character of sensations. The move is actually reminiscent of Descartes's attempt to refute skepticism with the aid of a nondeceiving God. Armed with that God, Descartes, too, wondered why we would have this natural inclination to believe in the existence of physical objects if our sensations were caused in some radically different way. But the question, of course, is whether any of this should be of interest to the skeptic. Even if we grant the rather controversial claim that sensations are intentional states (as opposed to being accompanied by intentional states), the skeptic is wondering whether anything like the intentional objects exist. One might as well argue that we do believe that physical objects exist and the most natural explanation of such a belief is that the belief is true. There is the view called epistemic conservatism that maintains that mere belief in a proposition does give that proposition initial epistemic credibility. I think the view has been successfully attacked,[22] but it will suffice for our present purposes to point out that if a refutation of skepticism depends on this move, the key to refuting skepticism is not reasoning to the best explanation, but epistemic conservatism. The fundamental principle on which this Keynesian will be relying is the principle that S's believing P makes P prima facie likely.

It is not clear, then, that even if we accept a principle like "Simpler explanations are more likely to be true than complex explanations," our commonsense beliefs will better satisfy criteria of simplicity than alternative explanations. But a skeptic who is prepared to raise difficulties with commonsense beliefs about the physical world and the past is hardly likely to simply grant you the hypothesis that simpler explanations are, ceteris paribus, more likely to be true. The Keynesian, you recall, is committed to the view that the simplicity hypothesis, if true, is necessarily true and knowable a priori. But the skeptic will almost certainly object that if simpler explanations are true more often than complex explanations, this is a *contingent* fact knowable only a posteriori. If we are to use that hypothesis in reaching conclusions, our internalism requires us to be justified in believing it. But what evidence can we adduce in support of this contingent hypothesis? The problem is similar to the problem we have already discussed with respect to justifying belief in the hypothesis that most events have explanations. Are we inductively justified in believing that simpler explanations are more likely to be true than complex explanations? If we are, is there any hope of justifiably believing the premises of this inductive argument without a prior and independent solution to the problem of justifying beliefs about the past based on memory experience? And if we do have inductive justification for believing both that everything has a cause and that simpler causal explanations are more likely to be true, then our so-called reasoning to the best explanation will not be a fundamental process of inference. It will be subsumed under inductive reasoning.

Now the *Keynesian*, determined to make a go of reasoning to the best explanation, should insist at this point that the above skeptic is confusing a frequency sense of probability with the internal epistemic relation of making probable. It seems almost obvious that it is a contingent fact (if it is a fact at all) that simpler explanations turn out to be true more often than complex explanations. But, for all that, it might still be necessarily true that if an explanation is simpler it is more likely (in the Keynesian sense) to be true. The Keynesian, like virtually any internalist or *sophisticated* externalist, will divorce the concept of epistemically rational beliefs from the concept of beliefs that are usually true. So we are left with the following question: Is it a contingent fact that an explanation's being simple increases the (epistemic) probability of its being true? Or is this principle only a secondary epistemic principle, for example, a principle that is justifiable only inductively? Once one realizes what one *needs* in order to use reasoning to the best explanation in the defense of common sense, one may well find more attractive the more straightforward Keynesian attempts to refute skepticism discussed below.

I should add to the above discussion that there is at least one very plausible a priori appeal to simplicity that derives from the epistemic counterpart of principles of the probability calculus. Suppose that I start out with two hypotheses, *H1* and *H2*, and that relative to my evidence *E*, *H1* and *H2* are equally likely. Suppose further that I acquire some additional evidence which requires me to either abandon *H2* or *add* to it an additional hypothesis *T1* while *H1* remains fine as it is. Clearly the conjunction (*H2* and *T1*) seems to have a probability lower than the probability of *H1*. But all this presupposes that we started out with *H1* and *H2* having equal probability relative to the original evidence base. There is nothing in any of this that suggests that, other things being equal, simpler hypotheses are more likely to be true than complex hypotheses.

Simplicity is, of course, just one of many criteria that philosophers have advanced for evaluating explanations. We should also be concerned, the argument goes, with how much a given hypothesis can explain, and how well we can assimilate the explanation to, other already familiar kinds of explanation.[23] As was true of simplicity, however, it is not clear that such criteria will favor the hypotheses of common sense. Explanations involving minds as the causes of our sensations, for example, seem to explain just as much as explanations involving physical objects, and as Berkeley himself pointed out, if we are relying on analogy we at least have experience of our own minds producing mental states to rely on in understanding how some other mind might be able to produce our sensory images.

Of course, just as some philosophers praise explanations that explain a great deal, other philosophers complain that skeptical scenarios explain too much. *Everything* that happens can be explained in terms of some alteration in the complex mind of a very powerful being. Of what use is such an explanation in making predictions? The answer is simply that it is as much use as any alternative explanation. As Ramsey pointed out long ago, we can substitute for the theoretical terms dummy placeholders for unknown causes and we will be none the worse off as far as predictive power. Since in the final analysis all we ever have to go on is what we find in the world we observe directly, any useful correlations that we find in that world can be mapped onto the unknown world of unknown causes and effects.

As was true of simplicity, we must ultimately come to grips with the modal status of the claim that comprehensive and familiar explanations are more likely to be true. The question once again is not whether it is a contingent fact that explanations satisfying these criteria are true more often than explanations that fail to satisfy the criteria. This can be granted by the Keynesian, who goes on to insist that it is nevertheless a necessary truth that explanations satisfying these criteria are more (epistemically) likely to be true. And the same question we are raising here can be raised with respect to any additional criteria philosophers might suggest for choosing between alternative explanations. The skeptic (with whom I am sympathetic on this score) is convinced that epistemic principles that rely on any of the aforementioned criteria for choosing among explanations are *secondary* epistemic principles. In order to rationally accept and employ such principles, *experience* must confirm that relying on them works out rather frequently. The skeptic, in other words, will insist on an *inductive* justification of the acceptability of such principles. The Keynesian, however, can bite the bullet and advance the respective probability claims as synthetic a priori principles.

If one could develop a plausible principle (or set of principles) that captures reasoning to the best explanation and that actually resolves the skeptical challenge in favor of common sense, it might make it more attractive to embrace the needed principles as necessary a priori truths. But as I tried to point out, even if it is necessarily true that events probably have explanations, that explanations are probably simple and comprehensive, and that explanations probably have similar underlying structures (favoring familiar over unfamiliar explanations), it is not clear that we can employ these principles to choose the hypotheses of common sense over some of the famous skeptical scenarios. Again, if we think about the problem of perception, it seems to me that skeptical hypotheses that posit minds rather than physical objects as the causes of our sensations satisfy the above criteria *better* than the conclusions of common sense. If one is going to be an inferential internalist with a Keynesian concept of probability, one might as well take full advantage of the view and address the skeptical challenge in more straightforward ways guaranteed to achieve antiskeptical conclusions.

The Principle of Indifference

Before looking at epistemic principles tailor-made to resolve skeptical problems, we might briefly examine another potentially useful and much more plausible candidate for an *a priori* epistemic principle—the principle of indifference. Crudely stated, the principle of indifference tells us that if we have a list of competing hypotheses that exhaust the alternatives and we have no evidence that bears on the likelihood of any of them being true, then we should assign them all an equal probability. One might hope to use such a principle, together with the relatively uncontroversial appeals to simplicity sketched earlier, to get *somewhere* starting from *nowhere*. The great attraction of a principle of indifference for someone wanting to battle the skeptic with a clear conscience is that one can begin using the principle from a position of complete ignorance.

In fact, it seems to me that one ought to be able to develop some plausible version of the principle of indifference as one of the best candidates for a synthetic a priori principle of *epistemic* probability. The technical problems with doing so, however, are legendary. One of the problems concerns the ''counting'' of alternatives. It is either raining outside now or it is not raining outside now. Suppose that I have no evidence on which to rely one way or the other (I have forgotten whatever evidence might have been inductively relevant to reaching a conclusion). Shall I then conclude that relative to my ignorance the hypotheses are equally likely? Or should I reflect on the fact that I can ''divide'' the hypothesis that it is not raining into the hypotheses that it is sunny, that it is snowing, and that it is hailing, giving each a probability of one in four (for me)? But, of course, I can also subdivide the hypothesis that it is raining into the hypotheses that it is raining cats and dogs, that it is drizzling, and so on. It is tempting to suppose that in employing a principle of indifference one must invoke something like the concept of a genuine atomic state of affairs. One starts using indifference only after one has ''divided'' each possible state of affairs into its simplest components.

The difficulty of employing a principle of indifference without such constraints can also be easily shown by means of paradoxes that can arise. Russell, for example, asked us to consider the situation in which you know that a car has traveled one mile and has taken somewhere between a minute and two minutes to complete the journey. We also know, therefore, that the car has traveled at a speed somewhere between 30 and 60 mph. Since we know nothing else about the situation, it is extremely tempting to suppose that there is a fifty/fifty chance that the car traveled between 30 and 45 mph and a fifty/fifty chance that the car took between a minute and a half and two minutes to make the journey. But if the car traveled at 45 mph it would have taken less than a minute and a half to travel the mile. If we assume that probability is transitive over known entailment, we can generate a contradiction.

If it is true that a principle of indifference can only be used over a set of atomic states of affairs that constitute exclusive and exhaustive alternatives, it becomes less obvious how one can use it coupled with other principles of reasoning to move from a position of ignorance to commonsense conclusions.

Principles Guaranteed to Deliver

If you can convince yourself that Keynes was right and that there is a sui generis, unanalyzable relation of making probable holding between propositions, and you are committed to the rejection of skepticism, the dialectically most attractive position is surely to meet the skeptical challenge head on by ''discovering'' epistemic principles that bridge specifically the problematic gaps. In his writings over the past thirty years, Chisholm has done something similar with his epistemic principles. Chisholm, however, takes the fundamental epistemic concept to be the relational concept of a proposition being more reasonable to believe than another proposition. Using this primitive to define a number of other epistemic concepts, Chisholm goes on to advance epistemic principles that solve (or partially solve) epistemic problems. He rejects skepticism about memory, for example, by putting forth the following epistemic principle:

If *S seems to remember* having been *F*, and it is epistemically in the clear
for him that he remembers having been *F*, then it is *beyond reasonable
doubt* for *S* that he remembers having been *F*.[24]

The epistemological problems of perception are solved with principles
that relate appearances to the positive epistemic status of beliefs about
the physical world based on those appearances (see especially Chisholm
1989, 65). Chisholm has always been completely unapologetic about
his method of arriving at epistemic principles. He makes it clear at the
outset that he will not accept a skeptical conclusion and that he will
make whatever adjustments are necessary in his epistemology to avoid
skepticism (Chisholm 1989, chap. 1).[25]

Although they will employ a different primitive, Keynesians can follow in Chisholm's footprints. If one is convinced that one can legitimately infer propositions about past events from the fact that one seems
to remember them having occurred, one can advance the epistemic principle that *S*'s seeming to remember *P* makes probable that *P*. The proposition that I seem to see something round and red makes probable the
proposition that there is something there round and red. If I can know
(noninferentially) that I seem to remember *P* and that I seem to see
something round and red, and I can be aware of these probabilistic
connections (noninferentially), then I can legitimately infer *P* and that
there is something round and red. The problem of epistemically rational
beliefs about the future will be solved with a principle or principles of
induction. We might start with Russell's principles:

> When a thing of a certain sort *A* has been found to be associated with a
> thing of a certain other sort *B*, and has never been found dissociated from
> a thing of the sort *B*, the greater the number of cases in which *A* and *B*
> have been associated, the greater is the probability that they will be associ-
> ated in a fresh case in which one of them is known to be present. . . . The
> greater the number of cases in which a thing of the sort *A* has been found
> associated with a thing of the sort *B*, the more probable it is (if no cases
> of failure of association are known) that *A* is always associated with *B*.
> (Russell 1959, 66–67)

If additional principles of probability are needed to resolve problems
about knowing other minds or theoretical entities, they are readily available. One figures out what one needs, and one gives it to oneself.[26] For
the Keynesian the battle with skepticism is won decisively and with
very little effort.

As I indicated earlier, the fact that a philosopher is committed to the a priori status of a principle asserting a probabilistic connection does not mean that we cannot attempt to argue *against* the principle. It will not do, however, to complain that the Keynesian has given us no argument *for* the assertions of probabilistic connection. The dialectical beauty of the view is that no argument should be expected from the Keynesian. If someone is committed to the view that there are probabilistic connections that can be known a priori, it would be question-begging to reject such a view on the grounds that its proponent was unwilling to argue for (provide evidence for) the truth of the principles. But are there arguments against the antiskeptic's claims of probability connections between memory experiences and the past or sensory experiences and the physical world? Can we find counterexamples to Russell's principles of induction?

If we start with the principle of induction, the answer would seem to be obviously yes. Has Goodman not conclusively established that the principle is absurd when applied to properties like ''grue'' (being green before *t* and blue after *t*)?[27] And do propositions describing memory experiences always make probable propositions about the past? Can an extremely forgetful person legitimately rely on memory?

One must remember, however, that *E* can make probable *P* even though there is some other body of evidence *F* such that (*E* and *F*) does not make probable *P*. It sounds initially absurd to suppose that just before *t* one can still take the fact that something has always been grue to make likely that it will continue to be grue, but perhaps the absurdity stems only from the fact that we know other things about grueness that immediately allow us to discount the probabilistic connection that does exist between the propositions. Certainly we can have good grounds for distrusting memory, but those grounds need not shake our belief that the proposition that we seem to remember *P* makes probable the proposition that *P*. When we have good grounds for distrusting memory, we know something else *E* such that the conjunction (I seem to remember *P* and *E*) does *not* make probable *P*.

But surely there are possible worlds in which memory, for example, is consistently misleading. In those worlds will propositions describing memory experiences still make probable propositions describing the past? There is no reason why the Keynesian's answer should not be affirmative. The epistemic relation of making probable is not defined in terms of, and in no way supervenes upon, *frequencies*. And, of course, it is precisely the intuition that memory is a source of rational belief about the past that leads so many to suppose that in a world in which memory (unbeknownst to us) constantly plays us false, we are still perfectly rational to believe what we do about the past based on our memories.

The clever Keynesian will not be defeated by counterexamples. The strategy is to take the inferences we are prephilosophically inclined to regard as legitimate (in all possible worlds) and formulate the corresponding principles asserting probabilistic connections between the available evidence and the conclusions we reach. Since we *choose* our principles so that they do accord with common sense, it is unlikely that common sense will refute them!

The Keynesian strategy described above for avoiding skepticism is, I suspect, dialectically impregnable. If I were a lawyer defending the antiskeptic, it is the argument I would use. For me, it has only one drawback. I cannot quite bring myself to believe that I am phenomenologically acquainted with this internal relation of making probable bridging the problematic gaps. I do not object to phenomenological appeals in philosophy. It should be clear from much of what I have argued in this book that I believe they are an indispensable part of philosophical method. But it should also go without saying that one must be scrupulously honest in one's use of phenomenological appeal. There is nothing wrong with introducing the sui generis, but one must be absolutely sure that one understands that about which one talks. One must be certain that one can isolate in thought a relation of making probable holding between propositions describing current memory states and propositions describing past events before one can in good conscience appeal to such a relation in order to avoid skepticism. And in the end, I strongly suspect that the probability relation that philosophers *do seek* in order to avoid skepticism concerning inferentially justified beliefs is an illusion.

The Search for an Illusion?

In the previous section we examined the search for probability connections, awareness of which can respond to the skeptical challenge. If one reaches the conclusion that one simply cannot find probability connections of the sort posited by the Keynesian, where do we stand? Shall we reconsider our rejection of externalism? Shall we conclude that we must reconsider the goal of philosophical epistemology? It seems to me that such a conclusion would be a genuine non sequitur. It is perfectly possible that when we do epistemology we are *looking* for epistemic principles that we can know without inference, when in fact there are no such principles. The concept of making probable as an internal relation holding between propositions might in the end be unintelligible. Perhaps it is one of those concepts that reveal massive confusion in human thought. But the fact that one cannot answer the question one asks does not make it reasonable to suppose that one is not asking that question. Still, if the search for probabilistic connections that will defeat skepticism is the search for an illusion, can we not provide some sort of explanation for such massive confusion?

On one reading of Hume, many of our judgments are not so much false as badly confused. They employ *spurious* concepts—concepts that seem to us unproblematic but involve confusing one thing with another. Our moral judgments, Hume argues, could not *really* be about anything but our sentiments.[28] But I suspect that Hume would be willing to admit that ordinary people and most philosophers, for that matter, have a tendency to *project* these feelings outward as if they were objective properties of external events. The concept of causation, according to Hume, involves similar tensions. Although Hume offered a relatively straightforward regularity analysis of causation, there are many passages in which he seems to admit that such an analysis fails to capture what people *think* they mean by causation. The concept of causal connection is the concept of *necessary* connection, he admits, and he goes on to acknowledge that there is no *necessary* connection involved in mere regularity.[29] For those who want *necessary* connection, Hume is prepared to offer an alternative ''subjective'' analysis of causation. *X* is the cause of *Y,* on this subjective analysis, when in addition to the presence of spatial and temporal contiguity, the idea of *X* determines the mind to form the idea of *Y,* the experience of *X* determines the mind to expect *Y.*[30] Hume knows that on reflection it should seem absurd to locate causal connection in the mind,[31] but he thinks that ordinary people confuse the internal compulsion of the mind with an external necessary connection.

Now I am not defending Hume's views about causation (whatever other defects it might have, the ''subjective'' definition relying on the *causal* concept of ''determination'' is blatantly circular). But the idea behind Hume's charge of confusion might be seized by the epistemologist who regards our thought about probabilistic connection as badly confused. Perhaps we do think of probabilistic connection as something external to us, something of which we could be in some sense aware, when all that is really present is the determination of the mind to expect one thing when we experience another. Do my sensory experiences make it likely that there is a physical object? Well, I certainly expect the object to be there in all its complexity when I have the relevant sensations. This expectation is so deeply ingrained that it involves no conscious process of inference. I do not think it is absurd to suggest that we might confuse our awareness of this *habit* with awareness of a pseudorelation holding between propositions.

I hasten to emphasize that even if the concept of an internal relation of making probable is unintelligible, I do *not* take this to be an argument for externalism in epistemology. I am convinced that the philosopher wants to satisfy philosophical curiosity on epistemological matters by ending a regress of justification through direct acquaintance with those aspects of the world that make thoughts true. In the case of inferential justification, the philosopher wants to end a regress involving inferential connections by finding probabilistic connections with which we can be directly acquainted. But if we employ frequency conceptions of probability, the desire to find this sort of foundation for epistemic principles will inevitably be frustrated. There is still hope for the epistemologist who wants to find a satisfying answer to epistemological questions, and that hope lies in discovering an alternative to a frequency conception of making probable. But it is nowhere written that everything a philosopher wants, a philosopher can have. It may be that some philosophical questions must remain unanswered, some philosophical curiosity must remain unsatisfied.

You can always turn to externalism, but it seems clear to me that an externalist metaepistemology is not going to provide philosophical help. You do not get a satisfying answer to a question by changing the question. You do not satisfy an existing goal by changing the goal. If the externalist in metaepistemology is suggesting that traditional ways of framing epistemological questions lead to a dead-end confrontation with skepticism, I am sympathetic. If the externalist is suggesting that we can *redefine* epistemic concepts in a way that makes them clear and intelligible, I am sympathetic. If the externalist suggests that these *redefined* concepts are perfectly useful in all sorts of contexts for all sorts of purposes, I am sympathetic. If the externalist maintains that we can recast traditional philosophical concerns employing externalist analyses of epistemic concepts, I think the externalist is wrong.

If one cannot find in thought sui generis probability relations holding between propositions, one may well despair of resolving skeptical problems within the framework of radical foundationalism. On the problem of justifying belief in propositions describing the external world, for example, one might begin to suspect that Hume was right when he suggested with respect to what man ends up believing that "nature has not left this to his choice." Perhaps nature has decided that the question of what to believe is too important "to be trusted to our uncertain reasoning and speculations."[32]

Hume's hypothesis, I suspect, is accepted by externalists, but they do not want its truth to cheat us out of knowledge and justified belief. And one of the most attractive features of most versions of externalism is that it makes it relatively easy for us to know what we think we know even if Hume is right. As long as "nature" (we now prefer to talk about evolution) has ensured that we respond to certain stimuli with correct representations of the world, we will know and have justified belief. Indeed, given externalist epistemologies, there is no difficulty in any creature or machine capable of representing reality achieving knowledge and justified belief.

It seems to me, however, that contemporary epistemology has too long let its philosophical analyses of epistemic terms be *driven* by the desire to avoid skepticism, by the desire to accommodate "commonsense intuitions" about what we know or are justified in believing. It is true that we describe ourselves as knowing a great many things. We also say that the dog knows that its master is home, the rat knows that it will get water when it hears the bell, and the salmon knows that it must get upstream to lay its eggs. But it seems clear to me that one need not take seriously our love of anthropomorphizing when analyzing the concepts of knowledge and justified belief that concern *philosophers.* If Wittgenstein and his followers did nothing else, they surely have successfully argued that terms like "know" are used in a wide variety of ways in a wide variety of contexts.[33] As philosophers, however, we can and should try to focus on the philosophically relevant use of epistemic terms. And the philosophically relevant epistemic concepts are those the satisfaction of which resolves philosophical curiosity and doubt.

I remain convinced that the kind of knowledge and rational belief that a philosopher wants, the kind of knowledge and rational belief that will resolve philosophical curiosity, involves the kind of direct confrontation with reality captured by the concept of direct acquaintance. And to secure philosophically satisfying justification for our all of our commonsense beliefs, the reality with which we are directly acquainted will have to include probabilistic connections. It may not be possible, however, to develop a coherent conception of probabilistic connection that allows us to be acquainted with the kind of fact that would make an epistemic principle true. Hume may have been right. It may not be possible to justify in a philosophically satisfying way much of what we unreflectively believe. If this should be true, we may still satisfy, of course, the externalist's criteria for knowledge and justified belief, and these criteria may even mark perfectly clear and useful distinctions between beliefs and the kinds of relations they bear to the world. Internalists will continue to feel, however, that the externalist has *redefined* fundamental epistemic questions so as to make them irrelevant to traditional philosophical concerns.

Notes

1. Always supposing, of course, that the hallucinator has no reason to think that the experience is hallucinatory.

2. Again, see Pollock 1987, chap. 4, for a survey and critical evaluation of a number of different concepts of probability. Pollock is primarily concerned with arguing that there is no nontrivial conception of probability relevant to the understanding of epistemic concepts. He overlooks in his discussion, however, the Keynesian concept of probability which, I shall argue, has the best chance of providing the internalist with an illuminating concept of inferential justification.

3. The obvious and much discussed problem is avoiding the charge of hasty generalization from a single source of correlations.

4. I have sketched a solution to the problem in an unpublished paper (Fumerton 1992), but I would not bet the farm on its being correct. The rough idea is that laws of nature are generalizations embedded in a hierarchy of generalizations that has more instances than any hierarchy with which it conflicts. Two generalizations conflict when they cannot both be true if they both have instances. Two hierarchies conflict when each entails a generalization that conflicts with a generalization entailed by the other.

5. See Russell 1948, pp. 401–02.

6. This approach to resolving epistemic problems is not new, of course. It is essentially Strawson's solution to the problem of induction (Strawson 1952, chap. 9).

7. See Pollock 1974 and Pollock 1987, chap. 7.

8. See again Pollock's excellent critique of the epistemic relevance of subjective conceptions of probability (Pollock 1987, chap. 4).

9. Foley 1987.

10. Keynes developed his important views about epistemic probability in Keynes 1921.

11. Russell 1959, p. 68.

12. Russell 1959, chap. 7

13. Russell 1921, p. 159.

14. See, again, Chisholm's objections to crude reliabilism (Chisholm 1989, chap. 8).

15. Though not the only way. I recently had the ''Monty Hall'' puzzle explained to me (by Richard Foley). The puzzle is named after the game show

host who asked contestants to choose between three doors. Behind one door was a valuable prize (say, $10,000); behind the other two doors were gag prizes (say, goats). After the contestant chooses (say the choice is door 1), Monty reveals a goat behind one of the other doors and gives contestants the opportunity to stay with their original pick or switch to the other unopened door. Like almost everyone else, I was convinced that it would make no difference, that the odds were now fifty/fifty that the original pick was correct. As it turns out, the odds are only 1 in 3 that the original pick was correct, and thus that there is a 2-in-3 chance of winning by switching (on a frequency interpretation of probability). Furthermore, one *should* be able to discover this a priori. As I said, you will find that virtually everyone reaches an incorrect a priori conclusion.

16. Again, see Harman 1970 for a proponent of the view that argument to the best explanation is a pervasive nondeductive method of reasoning under which under kinds of reasoning (for example, inductive reasoning) can be subsumed.

17. See Locke 1959, book 4, chap. 9, esp. pp. 328–29, and Russell 1959, pp. 22–23. See also Mackie 1969. Locke does not say in so many words that he is employing reasoning to the best explanation, but it seems to me that this is the most reasonable account of what he is doing.

18. Fumerton 1992.

19. My own (rather old-fashioned) view is that some version of the D-N model of explanation is still the right approach to understanding the form of an *ideal* explanation.

20. For a sophisticated defense of common sense as the better explanation, see Vogel 1992. Vogel does not, I think, attempt to resolve all of the problems I point to concerning reasoning to the best explanation. His arguments presuppose some of the background knowledge a more radical skeptic might not allow.

21. Moser 1989, pp. 88–105.

22. Foley 1987, chap. 7.

23. Paul Thagard 1978, 76–92, calls these criteria consilience and analogy.

24. Chisholm, 1989, p. 68. A proposition is beyond reasonable doubt for a person when believing it is more reasonable than withholding belief. A proposition is in the clear for a person when withholding belief in it is not more reasonable than believing it. Chisholm understands remembering *P* (as opposed to *seeming* to remember *P*) in such a way that remembering *P* entails *P*.

25. This perhaps overstates the case somewhat. The official justification for accepting Chisholm's epistemic principles is not that they avoid skepticism. It is clear, however, that his metaphilosophical commitment is to tinker with epistemological conclusions until it is obvious that his epistemology will deliver the conclusions of common sense.

26. And philosophers are getting more and more daring about what they are willing to give themselves. After noting, correctly, our extensive reliance on the testimony of others in forming our beliefs, Coady (1992) argues, in effect, that

we should include as a fundamental source of rational belief such testimony. Our reliance on testimony is not parasitic, according to Coady, upon any empirical investigation into its reliability.

27. Goodman 1955, chap. 3.

28. Hume 1888, pp. 468–69.

29. Hume 1888, pp. 163–64.

30. Hume 1888, p. 170.

31. Hume 1888, p. 167.

32. Hume 1888, p. 187.

33. For an excellent discussion of the wide variety of uses ''know'' has, see Wolgast 1977. Wolgast's work foreshadowed Williams's claim (1991) that there is no single concept of knowledge—knowledge claims must always be understood relative to a context.

References

Alston, William. 1988. "The Deontological Conception of Epistemic Justification." Pp. 257–300 in *Philosophical Perspectives 2: Epistemology*, ed. James
Toberlin. Atascadero, Calif.: Ridgeview Publishing Co.
———. 1989. "Internalist Externalism." Pp. 227–45 in *Epistemic Justification*. Ithaca: Cornell University Press.
———. 1993. *The Reliability of Sense Perception*. Ithaca: Cornell University
Press.
Armstrong, David. 1963. "Is Introspective Knowledge Incorrigible?" *Philosophical Review* 72: 417–32.
———. 1973. *Belief, Truth and Knowledge*. London: Cambridge University
Press.
Audi, Robert. 1993. *The Structure of Justification*. Cambridge: Cambridge Uni-
versity Press.
Ayer, A. J. 1952. *Language, Truth and Logic*. New York: Dover.
———. 1956. *The Problem of Knowledge*. Edinburgh: Penguin.
Barsolou, L. 1985. "Ideals, Central Tendency, and Frequency of Instantiation."
Journal of Experimental Psychology: Learning, Memory, and Cognition 11:
629–54.
Berkeley, George. 1954. *Three Dialogues Between Hylas and Philonous*, ed.
Colin Turbayne. Indianapolis: Bobbs-Merrill.
Blanshard, Brand. 1939. *The Nature of Thought*. London: Allen and Unwin.
BonJour, Laurence. 1985. *The Structure of Empirical Knowledge*. Cambridge:
Harvard University Press.
Butchvarov, Panayot. 1970. *The Concept of Knowledge*. Evanston, Ill.: North-
western University Press.
———. 1992. "Wittgenstein and Skepticism with Regard to the Senses." Pp.
110–33 in *Wittgenstein and Contemporary Philosophy*, eds. Teghrarian and
Serafini. Wakefield, N.H.: Longwood Academic.
———. 1995. *Skepticism With Regard to the Senses*. Unpublished manuscript.
Chisholm, R. M. 1957. *Perceiving*. Ithaca: Cornell University Press.
———. 1977. *Theory of Knowledge*. 2d ed. Englewood Cliffs, N.J.: Prentice-
Hall.

————. 1989. *Theory of Knowledge.* 3d ed. Englewood Cliffs, N.J.: Prentice-Hall.

Coady, C. A. J. 1992. *Testimony.* Oxford: Clarendon Press.

Curley, Edwin. 1978. *Descartes Against the Skeptics.* Cambridge: Harvard University Press.

Davidson, Donald. 1981. "A Coherence Theory of Truth and Knowledge." Pp. 423–33 in *Kant oder Hegel?*, ed. Dieter Heinrich. Stuttgart: Klett-Cotta Buchaudlang.

Descartes, Rene. 1960. *Discourse on Method and Meditations,* trans. Laurence Lafleur. Indianapolis: Bobbs-Merrill.

Dretske, Fred. 1969. *Seeing and Knowing.* London: Routledge and Kegan Paul.

————. 1970. "Epistemic Operators." *Journal of Philosophy* 67: 1003–13.

————. 1981. *Knowledge and the Flow of Information.* Cambridge, Mass.: MIT Press.

Feldman, Richard. 1988. "Epistemic Obligations." Pp. 236–56 in *Philosophical Perspectives 2: Epistemology,* ed. James Toberlin. Atascadero, Calif.: Ridgeview Publishing Co.

Firth, Roderick. 1959. "Chisholm and the Ethics of Belief." *Philosophical Review* 68: 493–506.

Foley, Richard. 1979. "Justified Inconsistent Beliefs." *American Philosophical Quarterly* 16: 247–58.

————. 1987. *The Theory of Epistemic Rationality.* Cambridge, Mass.: Harvard University Press.

————. 1990. "Fumerton's Puzzle." *Journal of Philosophical Research* 15: 109–13.

Fumerton, Richard. 1976. "Subjunctive Conditionals." *Philosophy of Science* 43: 523–38.

————. 1980. "Induction and Reasoning to the Best Explanation." *Philosophy of Science* 47: 589–600.

————. 1983. "The Paradox of Analysis." *Philosophy and Phenomenological Research* 43: 477–97.

————. 1985. *Metaphysical and Epistemological Problems of Perception.* Lincoln: University of Nebraska Press.

————. 1988. "The Internalism/Externalism Controversy." Pp. 443–60 in *Philosophical Perspectives 2: Epistemology,* ed. James Toberlin. Atascadero, Calif.: Ridgeview Publishing Co.

————. 1990. *Reason and Morality.* Ithaca: Cornell University Press.

————. 1992a. "Skepticism and Reasoning to the Best Explanation." Pp. 149–169 in *Philosophical Issues 2: Rationality in Epistemology,* ed. Enrique Villanueva. Atascadero, Calif.: Ridgeview Publishing Co.

————. 1992b. "A Regularity Theory of Law." Unpublished article.

Fumerton, Richard and Richard Foley. 1985. "Davidson's Theism." *Philosophical Studies* 48: 83–89.

Goldman, Alvin. 1979. "What is Justified Belief?" Pp. 1-23 in *Justification and Knowledge,* ed. George Pappas. Dordrecht: Reidel.

————. 1986. *Epistemology and Cognition*. Cambridge, Mass.: Harvard University Press.

————. 1988. "Strong and Weak Justification." Pp. 51–69 in *Philosophical Perspectives 2: Epistemology*, ed. James Toberlin. Atascadero, Calif: Ridgeview Publishing Co.

Goodman, Nelson. 1955. *Fact, Fiction and Forecast*. Indianapolis: Bobbs-Merrill.

————. 1978. *Ways of World Making*. Indianapolis: Hackett.

Harman, Gilbert. 1965. "The Inference to the Best Explanation." *Philosophical Review* 74: 88–95.

————. 1970. "Induction." Pp. 83–89 in *Induction, Acceptance, and Rational Belief*, ed. Marshall Swain. Dordrecht: Reidel.

Helm, Paul. 1994. *Belief Policies*. Cambridge, Mass.: Cambridge University Press.

Hockett, Eric. 1992. "Moser's Definition of Knowledge." Unpublished article.

Hume, David. 1888. *A Treatise of Human Nature*, ed. L. A. Selby-Bigge. London: Oxford University Press.

Keynes, John. 1921. *A Treatise on Probability*. London: Macmillan.

Kim, Jaegwon. 1988. "What is 'Naturalized Epistemology'?" Pp. 381–406 in *Philosophical Perspectives 2: Epistemology*, ed. James Toberlin. Atascadero, Calif.: Ridgeview Publishing Co.

Klein, Peter. 1981. *Certainty: A Refutation of Skepticism*. Minneapolis: University of Minnesota Press.

Lehrer, Keith. 1974. *Knowledge*. Oxford: Clarendon Press.

Locke, John. 1959. *An Essay Concerning Human Understanding*, ed. A. C. Fraser. New York: Dover.

Luper-Foy, Steven. 1985. "The Reliabilist Theory of Rational Belief." *The Monist* 68: 203-25.

Mackie, J. L. 1969. "What's Really Wrong with Phenomenalism." *Proceedings of the British Academy* 55: 113–27.

Moore, G. E. 1903. *Principia Ethica*. Cambridge: Cambridge University Press.

Moser, Paul. 1985. *Empirical Justification*. Dordrecht: Reidel.

————. 1989. *Knowledge and Evidence*. Cambridge: Cambridge University Press.

Nozick, Robert. 1981. *Philosophical Explanations*. Cambridge: Harvard University Press.

Pappas, George, ed. 1979. *Justification and Knowledge*. Dordrecht: Reidel.

Plantinga, Alvan. 1988. "Positive Epistemic Status and Proper Function." Pp. 1–50 in *Philosophical Perspectives 2: Epistemology*, ed. James Toberlin. Atascadero, Calif.: Ridgeview Publishing Co.

————. 1992. "Justification in the 20th Century." Pp. 43–78 in *Philosophical Issues 2: Rationality in Epistemology*, ed. Enrique Villanueva. Atascadero, Calif.: Ridgeview Publishing Co.

————. 1993a. *Warrant: The Current Debate*. Oxford: Oxford University Press.

————. 1993b. *Warrant and Proper Function.* Oxford: Oxford University Press.

Pollock, John. 1974. *Knowledge and Justification.* Princeton: Princeton University Press.

————. 1987. *Contemporary Theories of Knowledge.* Totowa, N.J.: Rowman and Littlefield.

Price, H. H. 1950. *Perception.* London: Methuen.

Putnam, Hilary. 1978. *Meaning and the Moral Sciences.* Boston: Routledge and Kegan Paul.

————. 1981. *Reason, Truth and History.* Cambridge: Cambridge University Press.

Quine, W. V. 1969. *Ontological Relativity and Other Essays.* New York: Columbia University Press.

Ramsey, William. 1992. "Prototypes and Conceptual Analysis." Unpublished article.

Raz, Joseph. 1992. "The Relevance of Coherence." *Boston University Law Review* 72: 273–321.

Rosch, E. 1975. "Cognitive Representation of Semantic Categories." *Journal of Experimental Psychology: General* 104: 192–233.

————. 1978. "Principles of Categorization." Pp. 27–48 in *Cognition and Categorization*, eds. Rosch and Lloyd. Hillsdale, Erlbaum.

Rosch, E., and C. Mervis. 1975. "Family Resemblances: Studies in the Internal Structure of Categories." *Cognitive Psychology* 8: 382–439.

Rosch, E., C. Simson, and R. S. Miller. 1976. "Structural Bases of Typicality Effects." *Journal of Experimental Psychology: Human Perception and Performance* 2: 491–502.

Rosch, E., and E. J. Shoben. 1983. "The Effect of Context on the Structure of Categories." *Cognitive Psychology* 15: 346–78.

Russell, Bertrand. 1921. *The Analysis of Mind.* London: Allen and Unwin.

————. 1926. *Our Knowledge of the External World.* London: Allen and Unwin.

————. 1948. *Human Knowledge: Its Scope and Limits.* New York: Simon and Schuster.

————. 1959. *The Problems of Philosophy.* Oxford: Oxford University Press.

Sartre, Jean-Paul. 1966. *Being and Nothingness.* New York: Simon and Schuster.

Sellars, Wilfred. 1963. *Science, Perception and Reality.* London: Routledge and Kegan Paul.

Smith, E. A., and D. L. Medlin. 1981. *Categories and Concepts.* Cambridge: Harvard University Press.

Sosa, Ernest. 1974. "How Do You Know?" *American Philosophical Quarterly* 11: pp. 113–22.

————. 1980. "The Foundations of Foundationalism." *Nous* 14. 547–64.

————. 1991. *Knowledge in Perspective.* Cambridge: Cambridge University Press.

Strawson, P. F. 1952. *Introduction to Logical Theory.* London: Methuen and Company.

Stroud, Barry. 1984. *The Significance of Philosophical Scepticism.* Oxford: Clarendon Press.

Swain, Marshall. 1970. *Induction, Acceptance, and Rational Belief.* Dordrecht: Reidel.

Thagard, Paul. 1978. ''The Best Explanation: Criteria for Theory Choice.'' *Journal of Philosophy* 75: 76–92.

Toberlin, James, ed. 1988. *Philosophical Perspectives 2: Epistemology.* Atascadero, Calif.: Ridgeview Publishing Company.

Villanueva, E, ed. 1992. *Philosophical Issues 2: Rationality in Epistemology.* Atascadero, Calif.: Ridgeview Publishing Company.

Vogel, Jonathan. 1992. Unpublished paper on argument to the best explanation.

Wolgast, Elizabeth. 1977. *Paradoxes of Knowledge.* Ithaca: Cornell University Press.

Williams, Michael. 1991. *Unnatural Doubts.* Oxford: Blackwell.

Index

About the Author

Richard Fumerton is professor of philosophy at the University of Iowa. He received his M.A. and Ph.D. degrees from Brown University and his B.A. from the University of Toronto. He is also the author of *Reason and Morality: A Defense of the Egocentric Perspective* and *Metaphysical and Epistemological Problems of Perception.*